GDT SPEEDSTER

from dream to reality

BY GENE D. DICKIRSON

Published by GDT Speedster, LLC, Plymouth, Michigan

First printing 2007

Printed in the United States of America
ISBN: 978-0-6151-6515-8 (soft cover/color interior)
ISBN: 978-0-6151-7326-9 (soft cover/black & white interior)

Library of Congress Control Number: 2007907210 (soft cover/color interior)
Library of Congress Control Number: 2007908455 (soft cover/black & white interior)

Compiled by Kristen Schmidt
Front Cover layout by Ken Ross
Front Cover drawings by Craig Sandvig

This book is dedicated to every member of the team who played a part in creating the vehicle. Also to their wives for their patience and understanding while their husbands worked on the car.

It is also dedicated to the friends and family members who provided words of encouragement throughout the entire process.

It is especially dedicated to my wife, Barbara, for her help and energy in making my dream come true.

.

Contents

1

Introduction

Designing and building the GDT Speedster was a labor of love. More than 13,000 cumulative hours were devoted to the effort by the team members with no assurance they would be compensated monetarily for their work.

Ever since I was a little boy I have had a passion for cars. My formative years were spent at my father's various gasoline stations. Early on I would watch how things were done. When I was about 13 years of age I began working as an attendant, car washer, lubrication technician, mechanic, janitor and any other chores that needed to be done. During those years I became well acquainted with the inner working of cars and trucks.

I began reading and collecting car magazines around the late 1950's. Hot Rod, Motor Trend, Rod & Custom, Car Craft and several others instilled the passion and desire to own and drive beautiful, customized cars. Like many other boys I built and customized plastic model cars. I built some model cars from scratch during my youth. The first was done around 1955. It was a carved wood body of a Ferrari- like open wheel racecar approximately 4 inches long with generic model airplane metal wheels and rubber tires.

Scratch built balsa model ca. 1955 - reproduction shown was built in 2006.

The second scratch built model was started around 1956 and used the guidelines published by the Fisher Body Craftsman Guild. Like many others, I gathered the materials for the famous contest but never completed a model. I kept the roughly shaped wood model for many years and finally completed it in 1982.

Wood model for Fisher Body Craftsman Guild in ca.1957, completed in 1982.

The third scratch built model was completed in 1960. It was a fiberglass bodied Model T roadster with swooping fenders. The 1:24 scale body "plug" was built using clay sculpted around a plastic model kit. Plaster of Paris was used for the female mold. I called the model "Iced T".

Scratch built fiberglass model ca.1960 based upon 1:24scale plastic kit called, "Iced T."

Along with my boy hood friends, brothers Bob and Bill Rodreick I also built a ride able "car" at around the age of 10. We built our four-wheel carts to be "drivable" i.e. they could be steered by a rope attached to the front axel support. Forward motion was achieved by pushing each other –" you push me then I'll push you".

Early push cart ca. 1954.

Later on I wrapped the rope around a broom handle with a steering wheel attached as a stab at "high technology." The cars were fabricated using boards, nails and 8" to 10" diameter solid rubber tired wheels and axels discarded from unknown original applications.

"High tech" upgrades include steering wheel and hood ca. 1955.

I built a model from scratch as a learning exercise beginning in 1996. The 1:8 scale model "plug" was sculpted in clay and plaster was used as the female mold. The open wheel design used the wheels from a 1:8 scale Monogram kit of a 1986 Corvette. The model was completed in December 1998 and served as a visual aid in discussing the future project with prospective suppliers and team members.

Scratch built fiberglass model made in 1998 as visual aid and reference for discussions.

During my high school years I aspired to become an automobile stylist with the thought that making sketches and renderings of futuristic cars would be the most exciting aspect of the automobile business. I enjoyed my high school art and mechanical drafting classes best of all. My father taught me to drive in the late 1950's in a Ford Model A coupe and the family's 1952 Buick.

Learning to drive in the Las Vegas desert ca. 1956.

During my high school era I had several different cars starting with a nice 1936 Ford 2 door sedan that I customized by adding a dropped front axel and converted the brakes from mechanical actuation to the safer hydraulic system. After "trading up" thru a 1927 Model T and a 1946 Ford coupe I ended up with a 1955 Plymouth. The black 2 door Plaza coupe was a head turner after I customized it by "leading in" the hood and trunk lid, lowering the front end to achieve a California rake, adding a late model Plymouth oval steering wheel and a floor shifter to replace the stock "three on the tree" system.

Customized 1955 Plymouth ca. 1960.

After high school in 1960 I applied to the Art Center School in Pasadena, CA with the goal to become a stylist. After the school officials reviewed my portfolio of car drawings they advise me to go to junior college for a couple of years to sharpen my drawing skills. After seeing other students' work at the school I felt it would take an enormous amount of training for me to get to their skill level. It was then that I decided to begin my career in the automobile business by becoming a draftsman rather than a stylist. After obtaining a degree in drafting technology from Vincennes University I was able to secure a position at Ford Motor Co. as a draftsman. I went to night school for 10 years to obtain a mechanical engineering degree.

During all of those years I always had the deep desire to design and build a real car from scratch. It was only after I retired from Ford Motor Co. in December 1998 that I had the time and financial resources to begin the project of fulfilling the life long dream of creating a one of a kind car- the GDT Speedster.

Rationale

The GDT Speedster was designed and built by a team of automotive professionals who love cars and who wanted to do a complete vehicle from scratch without the bureaucratic encumbrances of a large corporation. The project was considered to be an exercise in engineering and styling with "no holds barred"- in other words we could do what we wanted to appearance wise as long as the vehicle was functional and safe. It was also an engineering challenge because many projects like this one are started and very few are completed. Another goal was to complete the vehicle at a "reasonable" cost when compared to OEM produced one-off show cars.

Some people have asked, "Why did you do it? Why did you spend all that time and money to build a car when you could buy just about any car you wanted already built?" The answer is because it was a major personal challenge- like swimming the English Channel or climbing a mountain- the experts make it look easy after the project is completed but the doing is tough. The personal satisfaction is enormous after completing an endeavor that you put 100% into and knowing you did the very best you could every step in the process.

Name Selection

The team selected the vehicle's name, GDT Speedster. It stands for **G**ene **D**ickirson **T**eam Speedster. A speedster is the type of vehicle that typically has no top, no side windows and a raked back windshield intended for fun driving in good weather.

2

Team Formation

Gene Dickirson

For several years before beginning the project I knew the expertise and skills of other people would be needed in order to accomplish the task. Even though I had worked on the design and development of numerous automotive subsystems and components during my 35-year career at Ford Motor Company I was not skilled in the myriad of aspects that would be needed to design and fabricate a complete vehicle. Designing and building a complete vehicle is the dream of many "car" people and it was certainly mine. When I made up my mind to proceed with the project in 1998 I discussed it with my long time friend, car lover and fellow engineer Chuck Carlson.

Chuck Carlson

Chuck said he would like to participate on the project. Chuck retired from the Ford Motor Company at the end of 1999 after 36 years in the Product engineering Office. Chuck was an engineering manager with experience in climate control products and chassis systems. Chuck participated in essentially every aspect of the GDT Speedster project including design and fabrication. Chuck has a Bachelor of Science Degree in Mathematics from Wayne State University.

I then contacted Musa Azzouz and he agreed to help with the structural analysis on the project. Musa was a fellow professional engineer and he had been my professor for several courses at Lawrence Technological University. Musa retired after 42 years from the Ford Motor Company, Product Engineering Office where he spent most of his career as a body structures engineer. Musa provided design direction for the frame, underbody and restraint anchors to ensure the vehicle would be strong enough for its intended purpose. In addition to teaching at Lawrence Technological University Musa taught preparation courses for the Professional Engineers licensing exams on the subjects of Statics, Dynamics and Mechanics of Materials. Musa has a Masters of Science degree in Mechanical Engineering from Wayne State University.

Musa Azzouz

Larry Ronzi was a designer that I had worked with on a major project at Ford Motor Co. Larry was also a car lover and he agreed to join the team once I outlined the scope of the project. Larry Ronzi retired from the Ford Motor Company after 35 years in product design at the Design Center. Larry sculpted the 1:8 clay model working closely with Craig Sandvig on the overall design of the vehicle. Larry also sculpted the front portion of the hood of the full size foam mock up. Larry's skills include clay modeling, foam and paperboard mock up building, drafting, technical illustration, graphic arts, teaching engineers free hand sketching and industrial design. Larry has an Associate in Engineering degree in Mechanical Technology from Lawrence Technological University and a Bachelor in Fine Arts in Industrial Design from Wayne State University.

Larry Ronzi

Craig Sandvig worked directly next to Larry Ronzi at the Ford Design Center. Craig was not only a professional designer but also a hot rod builder. Craig joined the team as soon as Larry and I gave him the verbal plan. Craig Sandvig is a full time Ford Motor Company designer (stylist) working at the Design Center with more than 14 years experience. Craig had overall responsibility for the design (styling) of the exterior, interior and under hood of the vehicle. Craig also polished all of the aluminum parts and assisted with the fabrication of several components used on the vehicle. Craig has a Bachelors degree in Industrial Design from Wayne State University.

Craig Sandvig

Dave Maran and I had worked together for several years on various programs at Ford and then Visteon. Dave liked cars and agreed to join the team and help with any task especially those requiring his expertise in woodworking. Dave works in theVisteon Program Management Office where he has more than 32 years experience in product development and

Dave Maran

manufacturing. Dave provided numerous suggestions for the vehicle and fabricated several of the wood models/ welding fixtures. Dave has a Bachelor of Science degree in Business Administration from Lawrence Technological University.

James Wilber is a relative and an energetic engineer who likes cars. James is a Visteon design engineer on instrument clusters. James joined the team to help with the instruments and wiring. James assisted with the design and fabrication of the wiring system and instrument panel gages and A/C registers. James also assisted with the initial set up of my computer and assisted me in resolving CAD file translation issues. James assisted with research of Internet domain services alternatives. James has a total of 6 years experience in product engineering including detail design and customer interface of electromechanical components. James has a Bachelor of Science degree in Mechanical Engineering from the University of Evansville.

James Wilber

Keith Rogalski joined the team in August 2002. Keith and I had worked together previously on a project that utilized non-contact scanning at Visteon. Keith provided assistance with non-contact digital scanning. Keith has a total of 26 years experience in product engineering automotive/aerospace technologies. Keith has degrees from Henry Ford Community College, University of Detroit Mercy in Mechanical Engineering and Central Michigan University in the Automotive Vehicle Design program.

Keith Rogalski

Lawrence Conger was recommended by Keith Rogalski based upon their working relationship on Visteon programs. Lawrence joined the team in August 2002. He created the beautiful exterior body surfaces and the instrument panel surface for the GDT Speedster using ICEM/Surf, a powerful CAD tool. Lawrence is the owner of Digital Design, Inc. with 20 years of surface design experience in automotive, marine, recreational vehicle, agricultural and consumer products. His clients include more than 40 companies in 11 countries. Lawrence has a Bachelor of Science degree in Industrial Engineering/Computer Science from Central Michigan University.

Larry Conger

The first team meeting was held on 6 February 2000.

Team Status

The team is still intact and ready to begin a new project as this book goes to press in December 2007. Every member of the team expressed a great deal of satisfaction with participating in the 6-year project. Even though the financial reward was less than we had hoped for, every other aspect of the project was emotionally rewarding. Some of the members have moved away from South East Michigan but they have indicated they would like to participate on the next project in some way. This will be relatively easy to do with today's electronic technology.

Weekly status reports are still being published to keep the team members fully informed with any news related to the first vehicle or the next one. The team is a solid group of professionals with a combined automotive experience of more than 250 years.

3

Designing and Building It: The Process

The process used to design, develop and fabricate the GDT Speedster officially began on 6 February 2000 when the team held its first meeting. The first few meetings were held on Sunday afternoons at my office and workshop at Pine Trail Court in Plymouth, Michigan. Due to schedules it was more convenient for the majority of the team members to meet near their regular work offices so the meetings were held during lunch hours at the Beefcarver restaurant, initially in Allen Park and later in Dearborn, Michigan. Meetings were held approximately every two weeks. About three years later, after the major components of the vehicle were assembled the meetings were again held primarily at Pine Trail Court.

In general most of the work was done at Pine Trail Court however major chunks of the work was done at the homes of the team members when they were working on their area of expertise. For example Larry Ronzi did all of the clay modeling in his home basement studio. Larry Conger did all of the surfacing on his home-based computer. All of the welding, painting of the body panels and upholstery was done at the shops of the respective suppliers located in Metro Detroit.

A status report was written weekly and e mailed to the team members to keep them apprised of the progress on the project.

After the frame was completed a FINAL ASSEMBLY PROCESS list was compiled and maintained as work progressed. The sheets were intended to minimize "backtracking." The final list had 250 line items of completed actions.

Team celebrating the mid point of the project with the full size foam mock up.

Reviewing the cowl height proposal.

Team meetings were held regularly at a restaurant near the members' offices.

Reviewing the clay model at a team meeting.

Team meetings were also held at our workshop

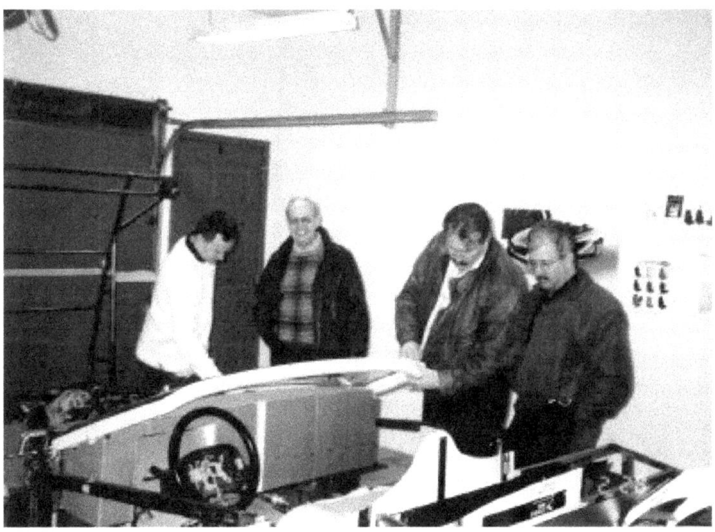

Process Comments

One of the major reasons the process of designing and building the GDT Speedster worked so well was because we did not set rigid time objectives for completing the myriad of tasks. Many of the team members were working full time at their "day jobs" and they all had families who needed a large share of their time.

Each member was eager to complete their individual portion of the project however we encouraged them to keep both their boss at the office and their boss at home happy before they worked on the vehicle.

4

Architecture Definition

The first step in the design process was to develop the basic assumptions for the vehicle. Several alternative architectures were considered including an open wheel car with a Spartan cockpit-sort of like a modern version of a hot rod roadster from the 1930's/1940's. The team soon agreed on the architecture. It was decided the vehicle would be a full fendered speedster with modern styling. Ted Kearnan, a retired Ford engineer who was involved during the first few months of the project was very helpful with developing the vehicle assumptions.

Craig Sandvig and Larry Ronzi began preparing theme sketches for the team to review. The sketches were the basis for selecting various styling elements used in the final design. The theme that evolved was fresh and it was intended to retain its freshness for many years to come.

14 March 2000

4 April 2000

6 April 2000

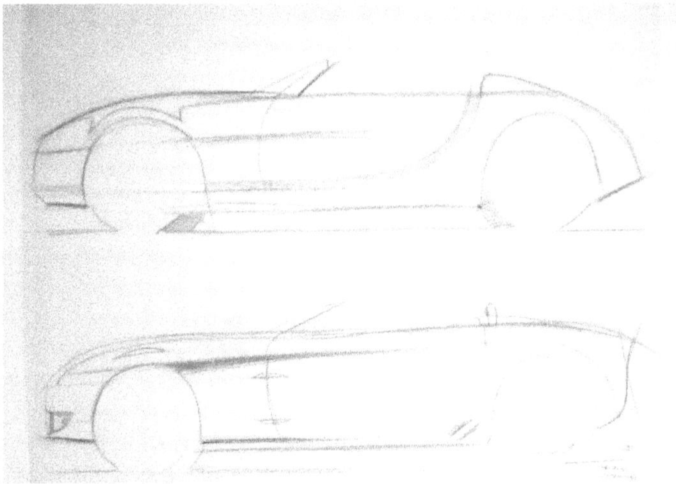

May 2000

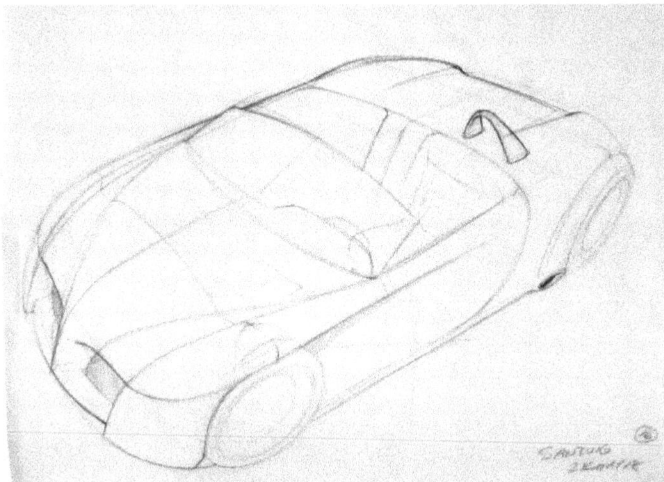

18 May 2000

18 May 2000

22 June 2000

22 June 2000

15 July 2000

15 July 2000

14 December 2000

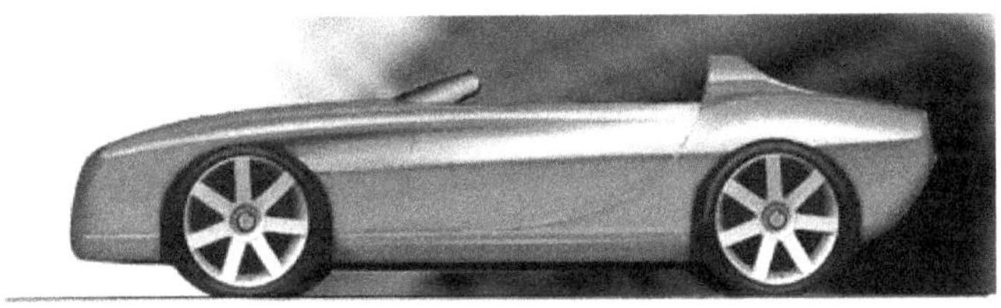

25 January 2001

15 March 2001

29 March 2001

15 April 2001

23 April 2001

Defining Hard Points

A critical element in the design of an automobile is the collection of "hard points" that establish the location of the driver and passenger with respect to their environment inside the vehicle. For example, the interior hard points include the location of the seat and steering wheel with respect to the doors, windshield, shift lever, floor, toe board and the space between the driver and passenger seats. The exterior hard points are the wheelbase, front and rear tread, ground clearance and turning radius among others.

We decided to use the hard points from the C4 as the starting point and vary only as necessary to achieve appearance objectives. The C4 hard points were determined from several sources including the data published in the Fall 1993 issue of Corvette Quarterly magazine, the cross section drawing in the March 1983 issue of Road & Track magazine and measurements taken from a 1/8 scale Monogram model car kit of the 1985 Corvette.

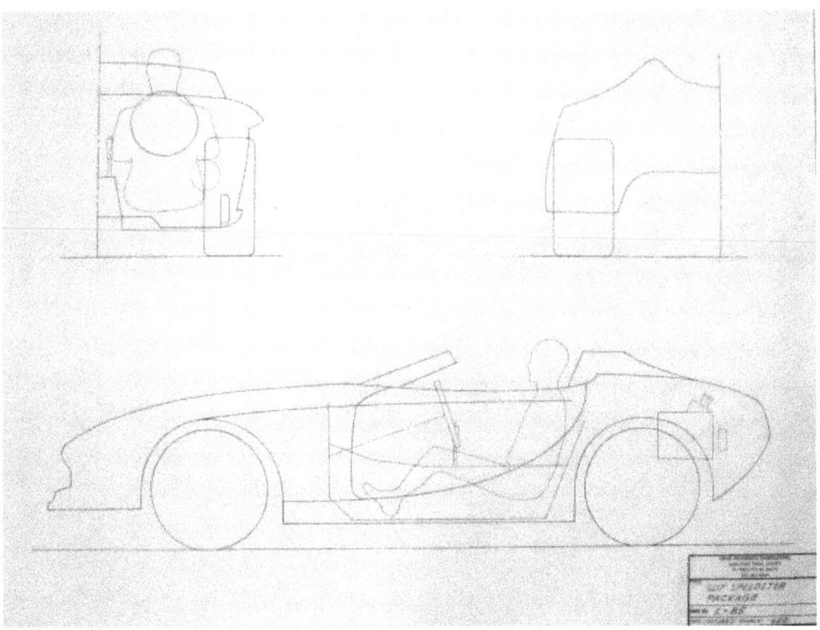

Package layout showing the major elements with respect to the driver.

Clay Model Initiated

An armature for a 1:8 scale clay model was fabricated after the basic hard points were established for driver location, shoulder room, cowl height, radiator package and wheelbase. An armature is a rigid structure for supporting modeling clay. The armature was securely attached to a surface plate. The Melamine surface plate had lines scribed on it 125 mm (4.92 inches) apart to facilitate accurate measurements between points on the clay model and the plate. The scale of 1:8 rather than something larger was chosen because we wanted the model to be portable between Larry Ronzi's home studio, our meetings, my studio and finally the laser scanning shop. Larger scale clay models would be too heavy to transport by hand. Our clay model, surface plate and carrying case weighed 40 pounds. Larry Ronzi "loaded" clay onto the armature and began sculpting the exterior surface of the body the week of 30 May 2000.

The wooden armature that was used to support the modeling clay and to locate some of the hard points such as the wheels and radiator.

5

Drive Train

It was decided early on to use the drive train from a 1984-1996 "C4" Corvette. The C4 drive train was selected for several reasons: 1. it is a well-engineered system, 2. it was fully developed by 1994, 3. service parts are readily available world wide, 4. the drive train is essentially "stand alone" in that the front suspension is connected to the engine that is connected to the transmission that is connected to the differential/rear suspension and therefore precisely establishes the wheelbase and drive shaft angle, 5. it produces world-class handling and top speed. The C4 drive train attaches to the chassis frame with only 12 bolts therefore making it relatively easy to install and remove the frame during the development phase.

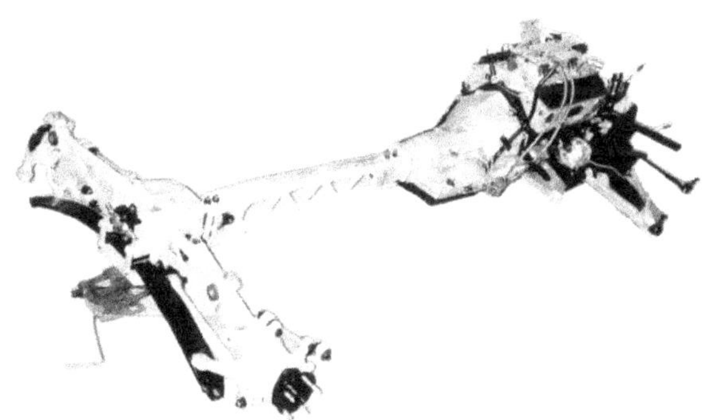

The C4 drive train is a well-engineered system that is attached to the frame with only 12 bolts.

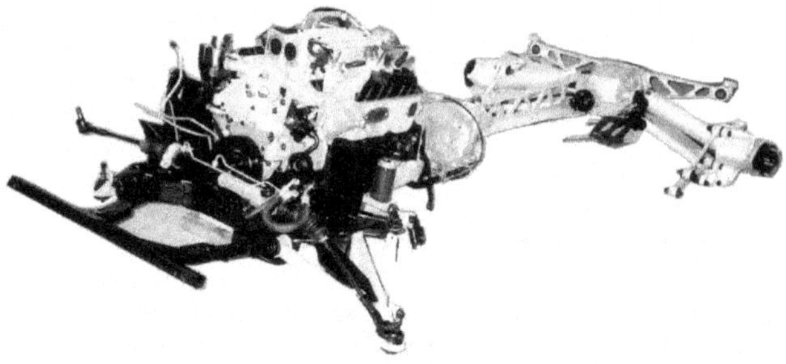

The C4 drive train precisely locates the front and rear wheels and controls the drive shaft angle.

6

Donor Vehicle

Soon after the architecture and drive train were selected a used 1994 Corvette was purchased on 9 August 2000. The "donor" vehicle had been in an accident that severely damaged the driver side of the body and "frame". The body panels, seats and interior trim had been removed by the recycling yard. The vehicle was placed upon jack stands and carefully leveled front to rear and side-to-side. Next a measuring "bridge" was constructed. . The bridge permitted accurate measurements of the vehicle front to rear and side to side with respect to "O-O" lines on the bridge side rail. Heights were measured from the concrete garage floor.

The donor vehicle was "reverse engineered" using the following items as cross-references: the 1994 Corvette shop manual, a body shop frame straightening chart by Mitchell, the sectional view from the Road & Track magazine and the Monogram 1:8 scale model kit.

Hundreds of measurements were taken and recorded on freehand sketches of numerous components and sub systems.

The dimensioned sketches were the basis for designing the new vehicle on full size vellum layouts. Points of interest from the donor vehicle sketches were "laid in" on the layouts then the new design was developed around the known points. Five full size layouts were drawn on 1067 mm (42 inch) rolls of vellum up to 4.3 m (14 feet) long showing the following views: side, plan- left side, plan- right side, front and rear. In addition, 79 full size layouts were drawn on "cut size" vellum 609 mm x 914 mm (24 inch x 36 inch).

Donor vehicle being driven off the trailer and into the garage.

Donor vehicle in fixture system.

Donor vehicle set up for measuring hundreds of points.

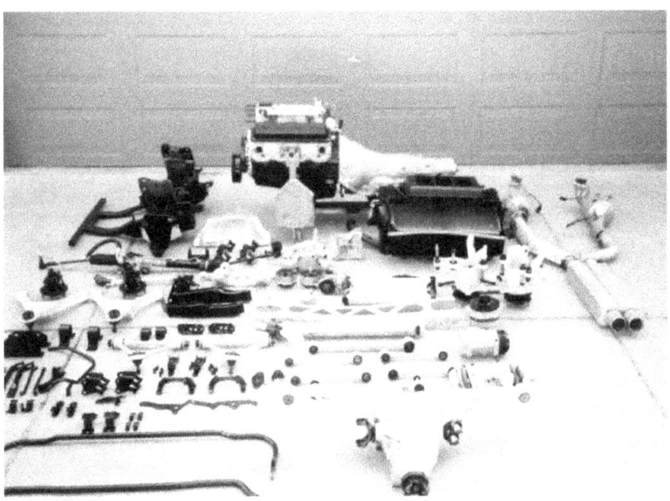

Components recycled and refurbished from the donor vehicle.

Reverse Engineering Comments

The ideal situation for reverse engineering a complete vehicle is to have the frame securely attached to a steel surface plate combined with a multiple axis measuring probe tied to a computer. The data can be stored in a spreadsheet and subsequently used to establish the "hard points" for the new design.

Because a steel surface plate was not readily available we opted to use the concrete garage floor as the horizontal O-O plane for measuring the donor vehicle. This was less than ideal because the garage floor was not absolutely flat. One section of the concrete sloped toward the garage door and this slope had to be accounted for as measurements were made in that region of the vehicle. The total drop of the floor from the rear of the donor vehicle to the front was 5 mm.

The side-to-side and front to rear measuring set ups were easier to control using long and short spirit levels and large perpendicular surfaces made of Melamine covered plywood.

7

Frame, Underbody, and Dash Panel

We did a preliminary study of using the "frame" from the donor vehicle. The assumption for the study was to repair the damage to the C4 structure and build new components above the side rails. The study included research of SAE technical papers and magazine articles on C4 road tests that indicated the C4 structure was not as rigid as some writers felt it should be. It was also felt that an all-new, scratch built frame would enhance the character of the vehicle and add to the engineering challenge. We decided to do an all- new, extra rigid frame.

Sketches were prepared of various concepts for the new frame. Rectangular ASTM A 500 steel tubing rather than round tubing was selected as the material for fabricating the frame. Rectangular tubing is much easier to cut and join than round. Round tubing requires precise "fish mouth" cuts in order to achieve good fits for welding. Some of the early frame concepts assumed the rectangular steel tubing would be formed by bending to achieve the desired shape. However after visiting with Bob Shetrone and his team at Progressive Automotive, Inc., a hot rod frame fabrication shop in Baltimore, Ohio in July 2000 it was decided to miter cut and weld the tubing rather than bend it to construct the frame. Progressive Automotive uses miter/weld joints rather than bends to produce its frames.

Aluminum was not considered for the frame because of its cost and extra degree of difficulty in fabricating. Aluminum is typically used in place of steel to save weight. Our target weight was the same as the stock 1994 Corvette therefore weight was not a concern and mitigated the need to use aluminum for the frame.

Musa Azzouz reviewed the frame concept proposals and offered suggestions on ways to achieve the desired strength. A preliminary frame design was prepared using CAD detail drawings of the frame components.

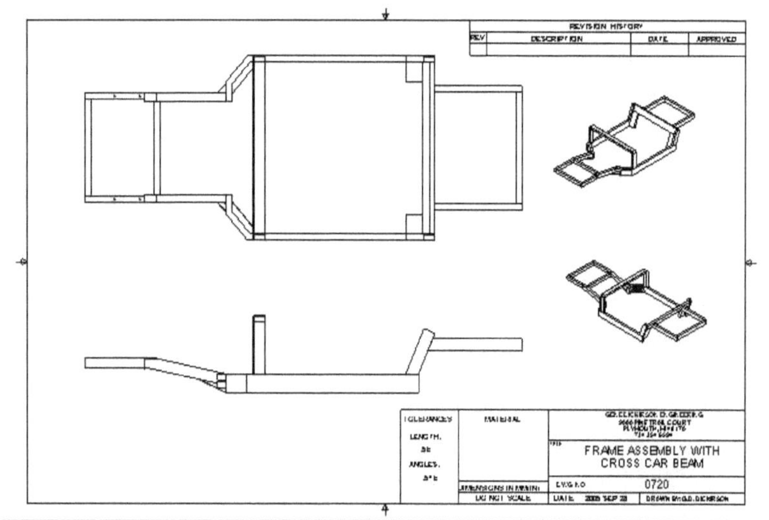

CAD drawing of the basic frame.

A 1:8 scale model of the frame was constructed using balsa wood and components from the Monogram plastic kit. Based upon an analysis of the 1:8 scale frame model the design of the frame was revised to improve fabrication feasibility and design strength.

Wood model in 1:8 scale used to develop the design of the frame.

A full-scale model of the driver side half of the frame was then constructed using 5 mm (3/16") thick foam core board and hot melt adhesive. Only the driver side half was modeled in order to reduce the fabrication time and to minimize the floor space required to study the mock up. Musa Azzouz reviewed the full size model and made recommendations for improving the design.

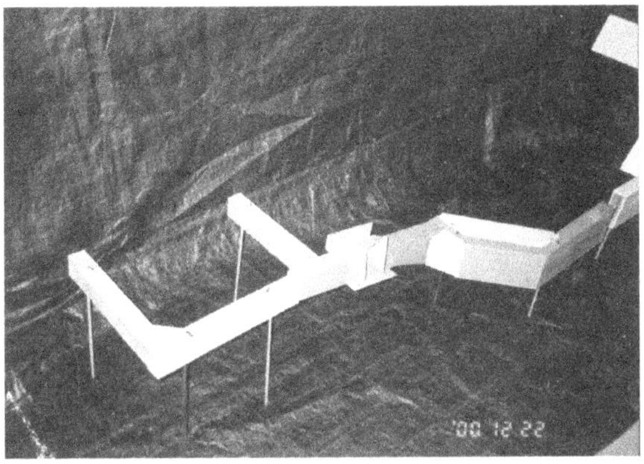

Full size mock up of the driver side of the frame made using foam core board and hot melt adhesive.

After the full size foam core board mock up of the frame was completed, analyzed and revised a second 1:8 scale model was built in wood. The model included the frame, engine, cooling system module, dash panel, cross car beam, differential, fuel tank, seats, wheel & tire envelope, transmission tunnel and cockpit floor. The model proved to be very useful in visualizing many aspects of the vehicle before functional components were fabricated.

After the frame design was completed on the full size layouts CAD detail drawings ("blue prints") were prepared for each component of the frame. The CAD detail drawings were "solid models" i.e. 3 dimensional representations of each part rather than conventional 2 dimensional. Solid models can be easily manipulated and "assembled" into a complete structure in the computer.

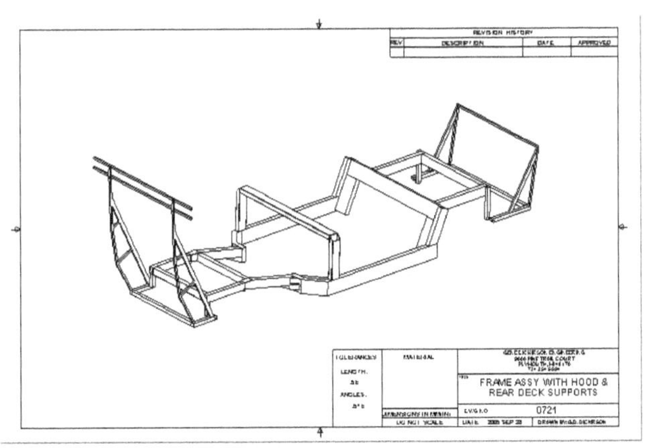

CAD drawing of the frame with the hood and rear deck supports in the open position.

The detail drawings of the steel frame components were sent to several different suppliers to obtain quotes. Weldtube Inc. in Dearborn, Michigan was selected to fabricate the tubular components. Weldtube made all of the precision cuts on the rectangular tubing to obtain the basic shapes. Federal Pipe & Supply in Detroit, Michigan was selected to fabricate several of the frame components from heavy gauge steel plate.

Fixtures for holding the frame components in place with respect to each other for welding were fabricated using a piece of 1219 mm x 1829 mm x 19 mm (4 foot x 6 foot x ¾inch) plywood, 51 mm x 102 mm (2 inch x 4 inch) blocks and machine screws driven into "cleat nuts" (three pronged T nuts). Several different fixture setups were made in order to weld the entire frame. The fixtures were fabricated in our workshop and transported to Starlight Welding in Redford, Michigan for final placement and welding. The Milewski's, Wally, Joe and Andy each had a hand in all of the 61 hours of welding done at Starlight.

The body/frame was removed from the donor vehicle drive train using a four wheeled movable crane especially fabricated for the task.

The new frame was fitted to the drive train and attached using the same 12 fastener locations as on the donor vehicle. The new vehicle frame and drive train was "rolled out" on 19 April 2001.

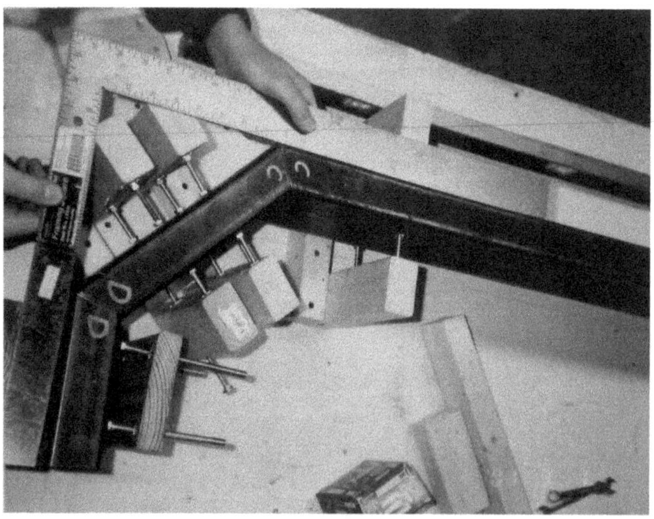

Fixtures for holding the tubular steel frame components in precise location prior to welding.

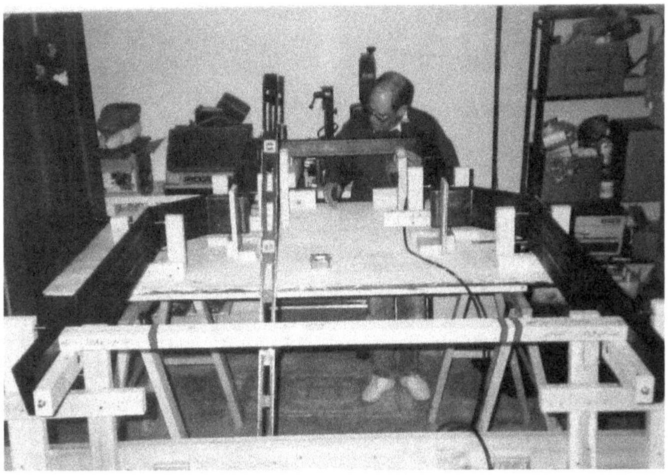

Frame components being positioned on the portable welding fixture.

Frame sub assemblies being secured for transporting to the welding shop.

Joe Milewski making the first weld on the vehicle frame on 22 February 2001.

Andy Milewski welding the frame while it was supported on the truck bed.

Fully welded frame being transported by a portable crane for fitting to the drive train.

First "roll out" of frame on drive train.

Frame Dimensional Check

The vehicle was towed to Starlight Welding on 21 August 2001 where the door pillars, cross car beam, transmission tunnel, dash panel, floor panels and the rear bulkhead panel were welded in place using continuous welds (versus skip welds which are less robust).

Frame being checked for dimensional accuracy vs. factory specifications.

The vehicle was later towed to Les Stanford Chevrolet in Dearborn, Michigan where the frame was checked for dimensional accuracy. It was found to be well within the factory specifications of plus or minus 4mm (.16 inches).

Door Pillars, Dash Panel, and Transmission Tunnel

The door pillars and cross car beam serve as a major structural element of the vehicle and they were fabricated from 51 mm x 102 mm (2 inch x 4 inch) rectangular steel tubing. They also establish the height of the cowl and the mounting surface for the door hinges. The location of the door hinges is critical because they directly affect the door opening geometry.

The cross car beam includes the door hinge pillars, steering column and brake pedal supports and the hood latches.

The dash panel, transmission tunnel, floor panels and rear storage compartment were developed on the vehicle and designed on full size layouts. The dash panel was mocked-up using 5 mm (.20 inch) Lauan plywood. Mock-ups of the “pockets” for the steering column mount and the brake vacuum booster mount were fabricated from 5 mm (.20 inch) Lauan plywood and steel drywall corner strips. The pockets were attached to the dash panel mock up using # 6 x 3/8 inch pan head sheet metal screws.

Dash panel, steering column and brake booster pockets being developed using plywood and drywall corners.

Completed dash panel with steering column and brake booster pockets.

The transmission tunnel top was also mocked up using 5 mm (.20 inch) Lauan and drywall corner strips. The transmission tunnel sides were mocked up using corrugated paperboard and masking tape.The transmission tunnel was developed as a set of flat rectangular and triangular shapes rather than a curved surface to make it easier to fabricate.

The floor panel and toe boards were located with the same relationship to the driver and passenger as in the donor vehicle i.e. same angle and same distance from the H Point. The H Point is the location of the occupants' hip joints while seated in the vehicle. The toe boards also are fitted with steel sheet metal heat shields on the underside to minimize the heat transfer from the catalytic converters to the passenger compartment and to shield the brake lines. After the design was finalized CAD details were prepared of the numerous components and sent out for quotes. Stewart Steel Specialties fabricated several of the sheet metal components and coordinated the laser cutting of the dash panel from heavy 13 gauge (2.3 mm- .089 in) steel plate.

Plywood fixtures to hold the transmission tunnel components in place for welding were fabricated by Dave Maran.

Transmission tunnel being developed using corrugated board and Luan plywood.

Completed transmission tunnel.

Frame Paint and Assembly

To prepare the frame for painting the steel structure was cleaned with an acid system, rinsed with water and dried immediately. A coat of PPG epoxy primer was applied. The joints in the floor panels and dash panel were sealed with Fusor 809 sealer. The dash panel was filled and smoothed. The frame was then painted with PPG DCC 5716 Concept in "Dark Shadow Grey Metallic" color.

Abrasive blasting using a mixture of glass beads and aluminum oxide cleaned the aluminum components recycled from the donor vehicle. The steel parts were cleaned then painted with Krylon Industrial "Rust Tough" black.

The frame was assembled to the drive train using the wheeled frame-lifting crane. The fuel and brake lines were installed and the tail pipes were loosely fitted.

Completed frame painted and ready for final assembly.

Frame Stiffness Comments

The objective of making the frame on the GDT Speedster stiffer than the C4 frame was clearly met when measured from a subjective viewpoint. The new frame resulted in a firm ride with little or no twist under all road surface-driving conditions.

It was noted that when a floor jack was placed under the side frame rail just in front of the rear wheel and the vehicle was jacked up the front wheel came off the ground to nearly the same height as the rear wheel. Furthermore the doors could be opened and closed with the vehicle raised on a jack.

For reference the frame weight was measured and found to be 249 kg (550 lb) with the cross car beam, dash panel and floor/ tunnel welded in place.

8

Windshield

The windshield is a major element in the design and functionality of a vehicle. Some research was conducted on custom fabricated glass windshields that would meet government standards. Even though plastic windshields are used on racecars and other off road vehicles and they are relatively easy to fabricate they are not government approved for use on passenger vehicles in the United States. The prices quoted for custom glass windshields ranged from $5,000 to $50,000 each. The other major concern with a custom glass windshield is service- if one is cracked or broken the replacement cost would be the same as the initial fabrication cost. Based upon these factors it was decided to use a windshield from a production vehicle that typically cost $100 to $200 each.

The windshield for the GDT Speedster was selected based upon its shape and size. In order to determine the best production windshield for the application the windshields on 28 different cars were measured on vehicles in used car lots and the dimensions were tabulated on a spreadsheet. Measurements were taken on the length of the centerline section of the windshield, the width at the bottom and the width at the top. Tom Laird at Laird's Glass and Trim in Plymouth, Michigan provided two cracked windshields, one from a 1965 Pontiac LeMans and one from a 2001 Jeep Cherokee for evaluation. The broken windshields were carefully measured and 1:8 scale drawings were made. The drawings were used to "cut sections" on the full size layouts and thereby permit the selection of the "optimum" windshield. Glassline in Plymouth, Michigan fabricated a mock up of a Mustang windshield in fiberglass. The Mustang windshield was rejected because it was too "tall" and the sides sloped in too much at the top. The windshield type used on the 2001 Jeep Cherokee was selected because it was the best fit for height and width and its upper corner radii were well proportioned.

Several incline angles were evaluated for the windshield. It was desired to have the angle as shallow as possible (close to horizontal) and still permit an acceptable viewing area between the vehicle hood and the block out mask at the top of the windshield. Evaluations of various angles were conducted on full size layouts using the nominal eye point location and the centerline section of the hood. The eye point location was determined on the layout by placing a seating mannequin on the layout. The mannequin's "H POINT" or hip point was placed on the seat H Point on the layout. The seat H Point was determined by placing a prototype Cerullo seat on the workshop floor and measuring the distance from the floor to the tops of the heads of two different people. Then knowing the distance from the floor to the top of the "average" head the H Point could be determined by using the mannequin on the layout. Dave Maran fabricated the seating mannequin from 5 mm (.20 inch) Lauan plywood. Larry Ronzi compiled the dimensions and patterns for the mannequin from published data. Dave named the mannequin "Biff".

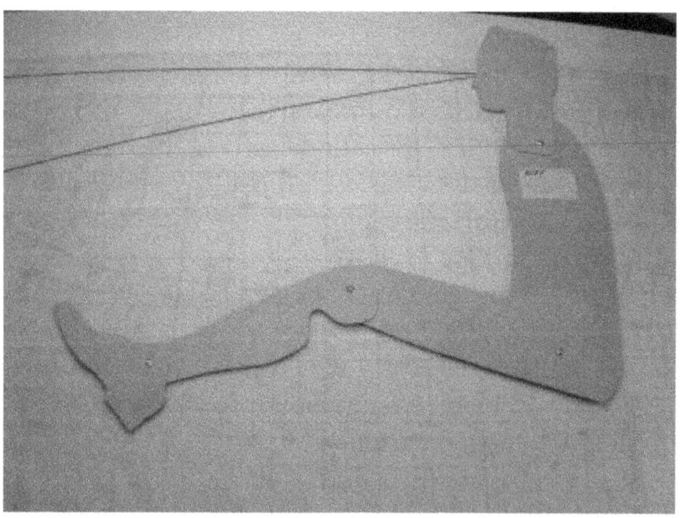

Biff, the wooden mannequin used on the full size layout to develop the windshield angle.

The team also evaluated the sight line alternatives in the vehicle using the cracked Cherokee windshield held at various angles using a temporary wooden support structure. A temporary seat, steering column and steering wheel were used for the in-vehicle evaluations. The final angle selected was 19 degrees from the horizontal.

Evaluating alternative windshield angles using temporary wood support fixtures.

Two different windshield support structures were designed and fabricated. Both used 19 mm (.75 inch) square steel tubing with 3 mm (.12 inch) wall thickness for the "A Pillars" or main side supports. The first windshield support structure had a header. It also had a fully molded cross section running around the "A Pillars" and the header. The first structure was evaluated on the vehicle and was judged by the team to be too bulky in appearance.

The first windshield support structure with a header was judged to be visually too bulky.

A second structure was designed and fabricated without a header and was judged to be ideal for achieving the desired "speedster" image of the vehicle.

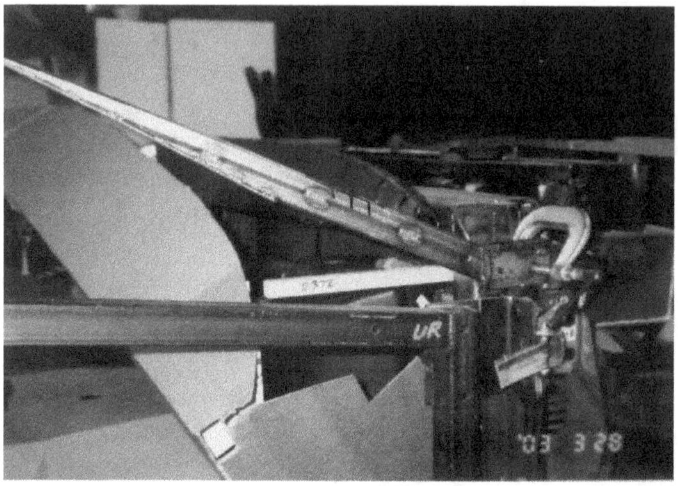

The final windshield support structure has minimal cross section yet is very strong.

In order to achieve a finished appearance it was desirable to cover the edge of the windshield where the two sheets of glass are bonded to the plastic core. Several alternatives were evaluated including extruded plastic moldings and lead separator strips used in stained glass artwork. It was desirable to minimize the cross section of the edge trim to enhance the appearance. The ideal solution would be a smooth edge that could be achieved with a thin, hard coating that would adhere to the glass. Eight different adhesives and coatings were tested on a broken windshield in order to identify the best material for the edge filler application. Super Fil epoxy by Poly Fiber was identified as the best material for the application. The windshield edge was coated with a layer of Super Fil epoxy putty, sanded, primed and painted.

Epoxy putty was applied, sanded and painted to finish the edge of the windshield.

The windshield block out pattern was modified to suit the GDT Speedster by placing the windshield in the vehicle and using masking tape to determine the line that would cover the leading edge of the instrument panel and the "A Pillars". Plastic masking tape (3M 218) and the PPG RS 21&22 two-part primer system were used to apply the new mask. Rick Walsh of Laird's Glass adhesively bonded the windshield to the support structure using the PPG RS polyurethane system. He also bonded the inside rear view mirror support button to the glass.

The windshield was adhesively bonded to the support structure.

The lower edge of the windshield is covered with a trim panel that also covers portions of the sides of the instrument panel. The trim panel is fabricated from vinyl ester resin and fiberglass cloth. The part was molded using the "splash" technique i.e. rough side out then smoothed. The lay up was done directly on the windshield glass covered with wax paper and on temporary forms made of chipboard and modeling clay. The panel is held in place using stainless steel #6 sheet metal screws at the sides and middle.

First step in fabricating the windshield lower trim molding using the "splash" method.

The finished windshield lower trim molding is held in place with stainless steel screws.

9

Body Panels

By 30 October 2001 Larry Ronzi had the 1:8 scale clay model of the body approximately 80% complete. Larry had been developing the model based upon some early "hard points" listed above (wheel base, shoulder room, etc.). It was now necessary to begin "cutting sections" thru the clay. The sections were necessary to verify clearance would exist between the body panels and the underlying structural/drive train components. More than 30 sections were "cut" thru the clay model.

To "cut" the sections the model was held firmly in place on a surface plate and lines were lightly scribed in the clay every 25 mm (0.98 inches) which was the equivalent of 200 mm (7.87 inches) on the full size vehicle. At selected scribed lines several points were measured. The vertical location of each point was measured from the surface plate (height). The side-to-side location of each point was measured from a "0" line scribed on the surface plate that was parallel to the longitudinal centerline of the model (width). The front to rear location of each point was measured from a "0" line scribed on the surface plate that was perpendicular to the longitudinal centerline and located approximately 72 mm (2.83 inches) behind the front axle.

The dimensions were tabulated on a spreadsheet. The dimensions were used to prepare 1:8 scale drawings on vellum of each section.

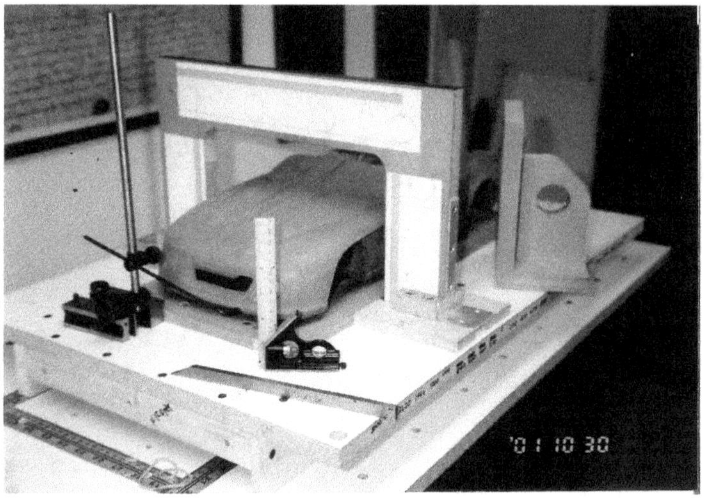

"Cutting sections" thru the clay model surface to determine clearances to the underlying components.

Full Size Mock-Up

It was decided that a full size mock up of the body surface would be highly desirable in order to obtain a more realistic view of the shape and size of the vehicle before going to the future steps of scanning/digitizing/surfacing/mold building/ molding. It would be extremely beneficial if the mock up could be viewed outdoors in sunlight from various distances and angles.

The first step in designing/fabricating the full size mock up was to modify the spreadsheet listing the 1:8 scale section dimensions. The spreadsheet was modified to calculate the full size dimensions of each point in each section by multiplying each dimension by 8.

A platform was designed to support the full size mock up. Because of limited floor space in the workshop the platform was fabricated in three sections. The width of each platform section had to be carefully designed to ensure that it could pass thru the workshop doors. In addition, the weight of each section had to be known in advance to ensure that it could be easily moved from the workshop to other viewing areas. The three platform sections were fabricated using 50 mm x 203 mm x 3657 mm (2 inch x 8 inch x 12 feet) pine boards with 5 mm (.20 inch) Lauan tops ("floors") fastened with drywall screws.

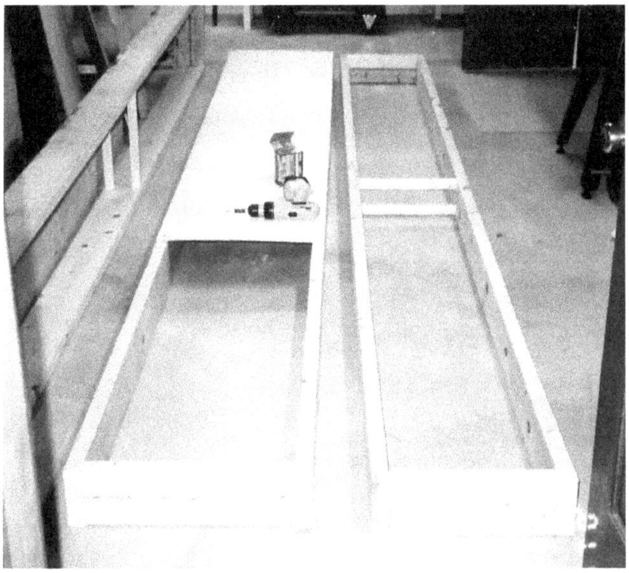

Three platform subassemblies for the base of the full size mock up.

The three platform subassemblies tacked together using furring strips and screws.

It was initially planned to use 11 mm (.44 inch) OSB (oriented strand board) for each of the 19 stations in the full size mock up. However after loading and hauling 453 kg (1,000 pounds) of the OSB to the workshop weight calculations were made. The calculations revealed the OSB was too heavy for the application. Lightweight Lauan 5 mm (.20 inch) thick was selected for the stations. The full size dimensions from the spreadsheet discussed above were used to place points ("dots") on the Lauan sheets. The points were joined using a felt tip pin and a 3 mm (.12 inch) diameter flexible plastic dowel rod. The stations were cut using a band saw and smoothed using a belt sander.

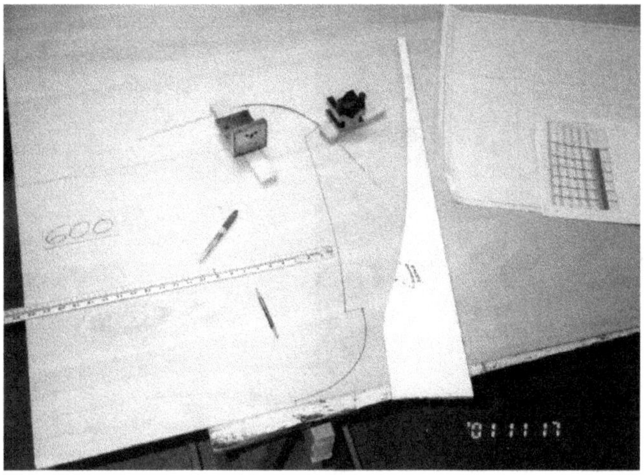

Transferring points from the clay model sections to the full size Lauan plywood station.

The three platform sections were "tacked together" using drywall screws and furring strips A centerline was scribed on the top (floor) of the middle platform front to rear. Lines perpendicular to the centerline were scribed on the tops of the platforms every 200 mm (7.9 inches) front to rear. Pine "furring blocks" 25 mm x 50 mm x 75 mm (1 inch x 2 inch x 3 inch) were attached to the platforms at each station line using hot melt adhesive. The 19 Lauan stations were attached to the platform furring blocks using dry wall screws Spacers were placed near the tops of the stations to prevent a "domino effect". The spaces between the Lauan stations were filled with pink polystyrene foam 38 mm (1.5 inch) thick using hot melt adhesive. The nose and tail sections and a windshield module were fabricated as separate pieces in order to make them portable. The foam was shaped using a Stanley Surform Shaver tool. A coat of Sheetrock Lightweight All Purpose Joint Compound was applied to the foam. The drywall joint compound was smoothed and then a coat of latex primer was applied. The topcoat was high gloss red latex paint.

The first stations being attached to the platform.

Foam blocks being attached to the stations using hot melt adhesive.

Foam being shaped with a Surform Shaver.

Stations for the nose section.

Nose section loaded with foam.

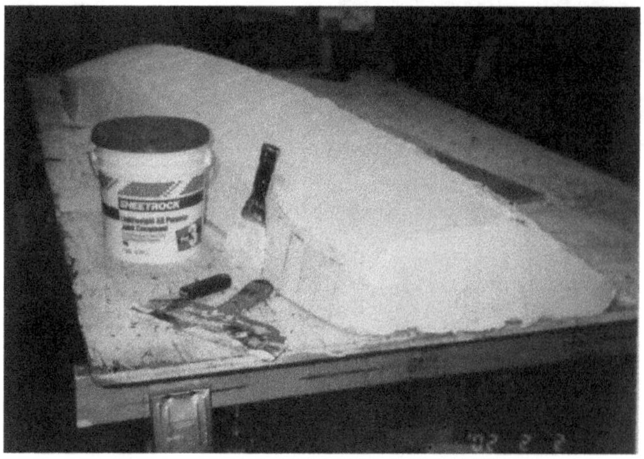

Drywall compound was used to fill the low spots-tail section shown.

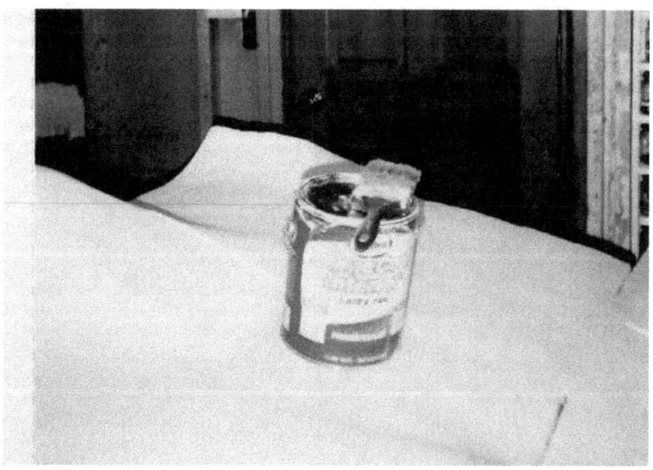

Latex primer was used to cover the drywall compound.

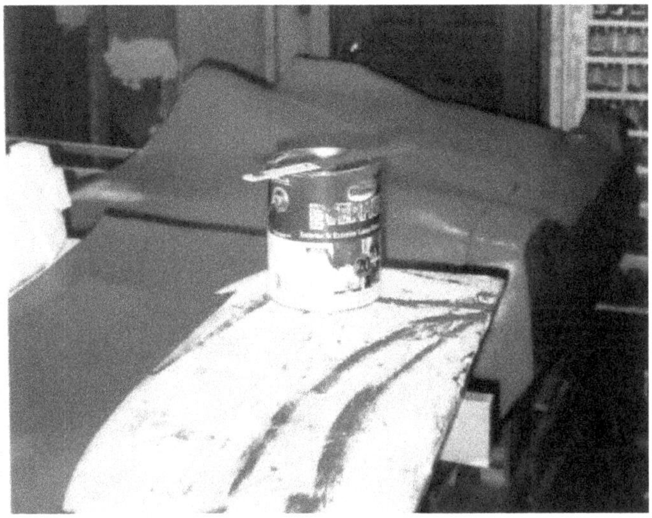

Glossy latex color paint was applied with a brush.

The wheels and tires were mocked up using paper pictures applied to Lauan plywood substrates. The headlamps were also mocked up using paper pictures glued to the foam.

On 15 February 2002 the five sections of the mock up were moved from the workshop out doors thru ice and snow then back inside to the adjacent family room. For the design review the joints between the five sections were covered with masking tape pre painted with the red latex paint. A design review was held with several of the team members present The design review of the mock up indicated several revisions would be needed to the clay model before proceeding to the "OK to TOOL" phase. The completed full size foam mock was reviewed in an outdoor setting as soon as the weather permitted on 31 March 2002.

Wheels and tires were mocked up using plywood, chipboard and photos.

Transporting driver side thru ice and snow from the workshop to the family room.

Final assembly of the full size mock up in the family room.

Indoor design review of the mock up revealed some areas that needed revision.

First look at the full size foam mock up in natural light.

Exploded view of the 8 major components of the full size mock up.

Tire Clearance to Fenders

One of the most obvious visual concerns identified with the mock up was the relationship between the front wheel/tire and the wheel well opening in the "front fender". The fender extended outward past the plane of the wheel/tire too far. This concern required that the front wheel tire jounce, rebound, left turn and right turn geometry be fully understood to ensure the tire did not strike the fender opening under extreme conditions. To determine one aspect of the geometry the vehicle was again positioned on jack stands and the left turn and right turn wheel/tire locations were carefully measured and the information was added to the layout.

To determine the jounce/rebound geometry it requires the front control arms be free to move i.e. the spring must be removed and it is difficult to gather accurate measurements without a surface plate or a computerized measuring machine. Rather than attempt to directly measure jounce/rebound it was decided to measure the relationship between the tire and fender opening on several production C4 Corvettes. A special measuring aid was fabricated and taken to the Les Stanford Chevrolet used car lot. The salesman on duty authorized us to measure several of the Corvettes on the lot.

The measurements were compiled and the average values were used to design the front tire/ fender gap on the new vehicle.

A jounce/rebound study was conducted on the rear wheel/tire. The vehicle was supported on jack stands and the rear spring was disconnected from the suspension system. The angle of the wheel was measured as the axle hub was raised then lowered to several different positions using a floor jack. This information was then added to the full size layout and used to position the rear fender opening with respect to the new wheels/tires.

A mock up of the wheel and tire envelope was used to develop the clearance to the fenders.

Wheel and tire looked tucked in too far in the original design.

Body surface was moved inward to be flush with the tire face to fix the concern.

Head Rest Development

The headrests were not fully developed in the full size mock up reviewed on 15 February 2002. The head rests were subsequently developed using "Biff", the seating mannequin in the vehicle. The head rests were developed so that they would be located approximately 150 mm (5.9 inches) behind the occupants' heads and provide adequate clearance to their shoulders. After the fiberglass rear deck panel was firmly attached to the vehicle and with a seat in place the area between the headrests looked "too heavy". It was decided to modify the space between the two headrests by moving the surface rearward approximately 150 mm (6 inches). This required cutting the fiberglass between the headrests and then bonding in 21 fiberglass segments to achieve the desired shape.

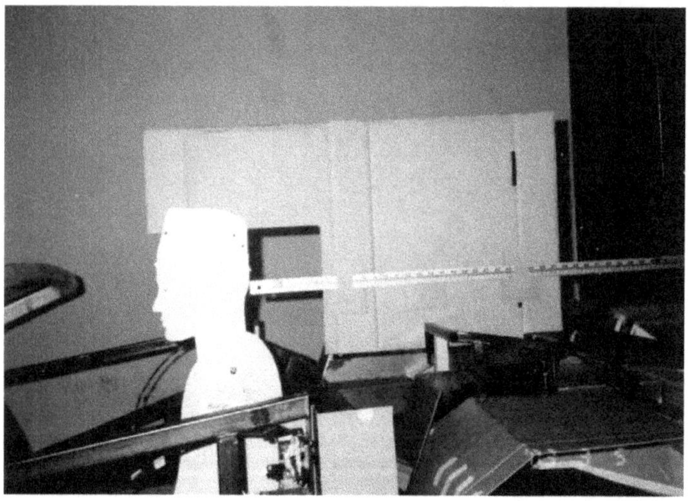

Headrest being developed using "Biff" in the vehicle.

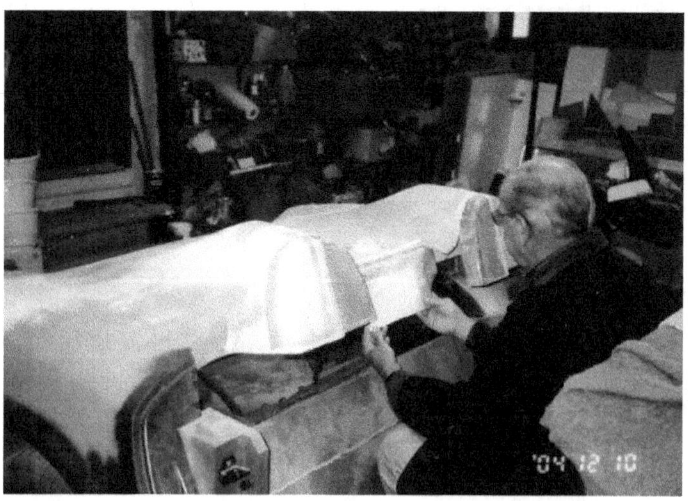

Revising the fiberglass to resolve the visual concern related to the headrests.

Door Cut Lines

The door cut lines represented a major challenge. They had to be attractive and functional. The doors had to clear the rocker panels and hood as they opened and they could not touch the rocker panels when they were fully opened. Several design iterations and models were built to achieve the desired results.

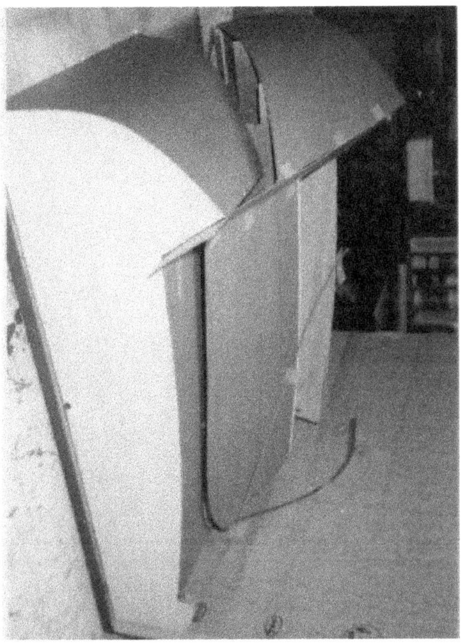

Side view of the mock up used to develop the complicated door cut line.

Front ¾ view of door cut line mock up.

Doorsills and Rocker Panels

The doorsills in the sides of the rocker panels were also a major design challenge. They have steps and pockets for the rear deck latches that were very intricate to design. A full-scale mock up was fabricated using corrugated paperboard and hot melt adhesive in order to fully visualize the surfaces during the design phase.

Mock up in corrugated board to develop the doorsill.

Completed doorsill.

Mock up used to develop the latch pocket and striker cover in the doorsill.

Completed latch pocket and striker cover.

Scanning and Digitizing the Clay Model

Larry Ronzi completed the 1:8 scale clay model on 17 October 2002 after incorporating all of the changes identified during the design review.

The clay model was then delivered to GKS Inspection Services, Inc. in Livonia, Michigan. The people at GKS scanned and digitized the model. The scanned data was in the form of a "point cloud" which consisted of the precise location of millions of points (dots) on the surface of the clay model. The data was recorded on a CD.

Clay model "sign off."

Clay model completed and ready for scanning.

Surfacing the Body

The CD of the point cloud scanned data was transferred to Larry Conger. Using a powerful CAD design tool (software package) called ICEM/Surf Larry converted the point cloud to a surface. He expanded the surface from 1:8 scale to full scale in his computer data file. He then smoothed the surface. Even though the clay model that Larry Ronzi sculpted looked and felt smooth it actually contained subtle imperfections that would become visible when expanded 8 times to full size. Next, Larry Conger made the right side of the body surface exactly symmetrical with the left side. In order to save time and enormous effort the 1:8 scale clay model was only finished on the left side, the front, the rear and the top. The right side of the clay was intentionally left rough. Larry Conger then developed the edges of each of the body panels. He also checked the door swing geometry to ensure all of the conditions discussed earlier were met.

After Craig Sandvig was satisfied that the body surfaces would look good in physical form Larry Conger e mailed the CAD file to Chris Mason at Method Industries in North Carolina.

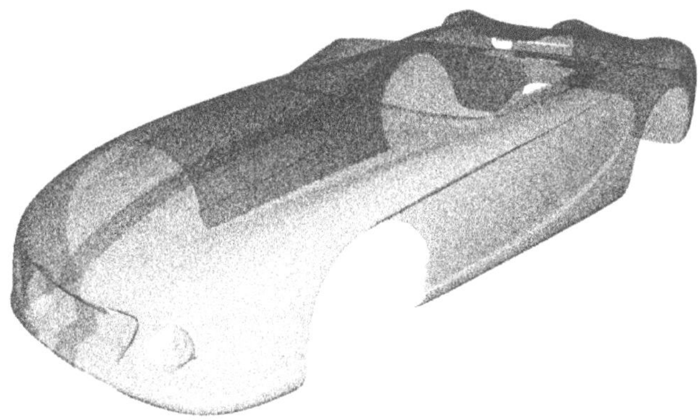

Point cloud comprised of millions of points gathered during digital scanning of the clay model.

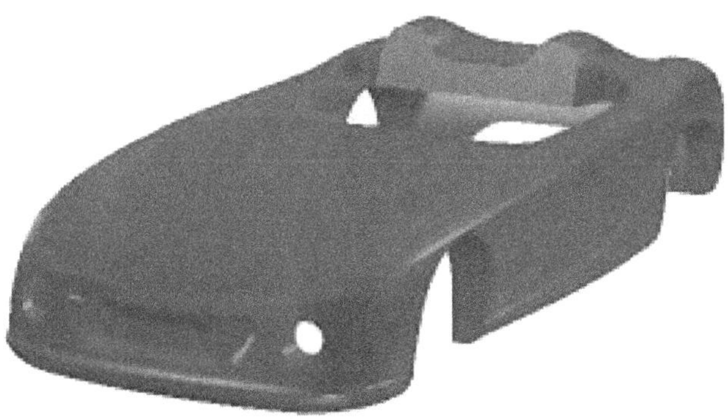

Body surface developed from the point cloud.

Molding Body Panels

Chris and his brother Andy then used the body surface CAD file to develop cutter paths for the 3 axis-milling machine they had developed for their shop. Cutter paths are the routes-up/down/left/right/fore/aft- the tip of the router must take to accurately remove material from each chunk of foam used for the molds. The molds for each part were made up of several chunks of 64 kg/m^3 (4 lb/ft^3) polyurethane foam that were adhesively bonded together.

Andy Mason (L) and brother Chris (R) with their milling machine.

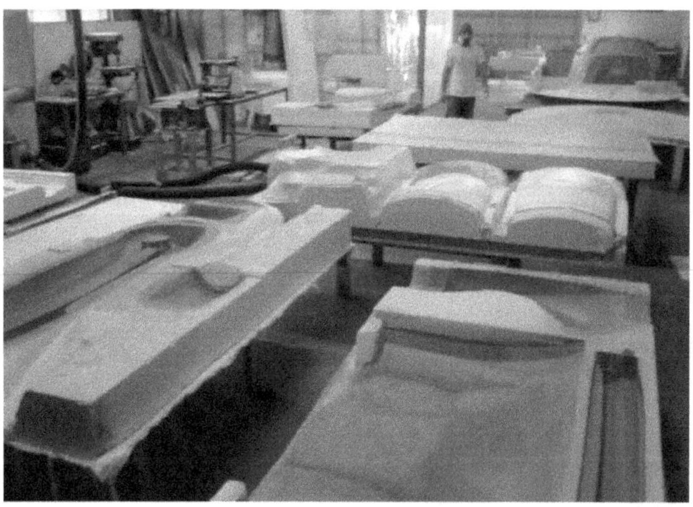

Molds were machined in large blocks of foam adhesively bonded together.

After machining, the "working cavity" of each mold segment was sprayed with a coat of Duratec resin that would serve as the "working surface" for the mold. After the resin hardened the working surface was smoothed then coated with a special "mold release" wax. The wax would prevent the molded fiberglass parts from sticking to the Duratec surface. The mold working cavities were sprayed with a "gel coat" that served as the top surface of the molded parts.

Mold for the driver half of the hood sprayed with filler and smoothed and ready for laying up of the fiberglass.

Up to 8 layers of fiberglass fabric, some Kevlar fabric and vinyl ester resin were then applied to each cavity to make the parts approximately 3 to 7 mm (.12 to .30 inches) thick. After the vinyl ester resin cured in a couple of days the parts could be removed from the molds. The molds were destroyed after the fiberglass parts were removed. The molded parts were then trimmed. They were then placed on fixtures and bonded to adjoining parts to make up the final body panels.

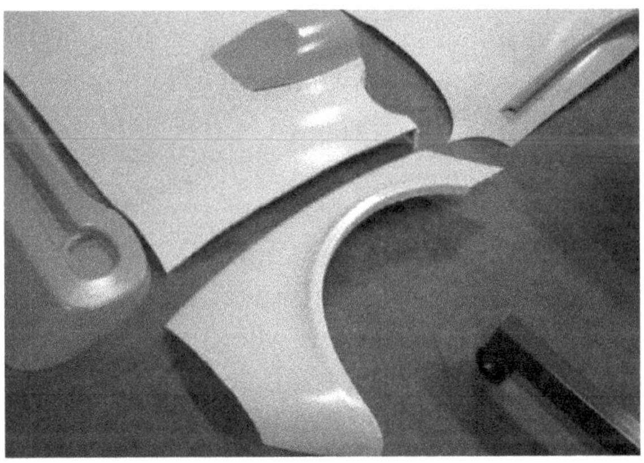

Molded fiberglass parts for rear deck and hood.

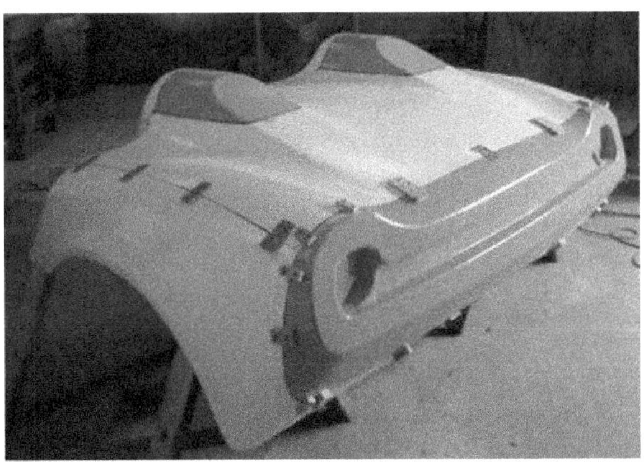

Components of rear deck tacked together for final joining.

The finished body panels were then packaged in special shipping containers we had designed. The containers were designed with input from the FedEx Freight people at their Romulus, Michigan depot. The panels were shipped from Chris and Andy's shop to the FedEx depot in Romulus where Chuck and I collected them and transported them to our work shop on a trailer.

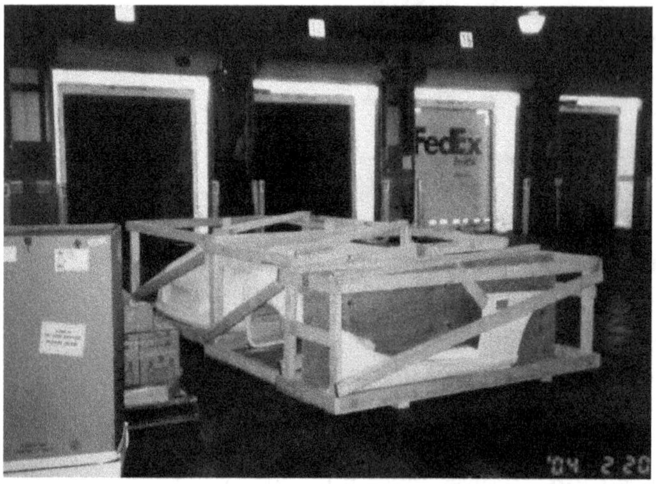

Body panels received at the FedEx depot in the special shipping containers custom designed for the parts.

The first two body panels were received from Chris and Andy Mason on 11 September 2003. They were the driver side door and rocker panel. They were temporarily attached to the vehicle and were judged to be excellent for finish. Fit would be evaluated later when the hood and rear deck were fabricated and installed.

A hurricane struck North Carolina about 20 September 2003 during the period the remainder of the body panels were being fabricated. The hurricane skipped over the workshop where the body panels were being fabricated however it inflicted extensive damage to the home of Chris and Andy Mason. They set up bunks and slept in their workshop until their home was repaired.

Door, Hood, and Rear Deck Support Structures

The door hinges were selected based upon their shape, size and robustness. They are the type used on the 1999 Ford F Series. The hinges had to be precisely located on the door pillars and the hinge pin centerlines had to be absolutely collinear in order for the doors to open and close properly. Door hinge technical specialists were consulted and they provided numerous design considerations. The door latches, strikers, interior handles and cables were also the type used on the 1999 Ford F Series. The door handles were located so that they would be at the ergonomically correct position when the occupant reached for them. Their location was developed by measuring the handle location on several production vehicles with respect to the driver's arm while setting in the vehicle seat. The measurements were averaged and the values used as the starting point for the handle location.

Door latch and handle system.

The door structural frames were fabricated from 38.1 mm (1.5 inch) square steel tubing with 3.1 mm (.12 inch) wall thickness. The door latches are mounted in fabricated five-sided steel "boxes" that are welded to the door structural members.

Door support structure and hinge system.

The hood support structure was designed using the full size layouts. The precise location of the upper engine components had to be shown on the layouts in order to ensure the structural components would clear them. To determine the location of the upper engine components a "measuring plane" was established above the engine in the new vehicle. Measurements were taken between the engine components and the measuring plane. The location data was transferred to the full size layout.

"Measuring plane" used to determine the precise location of engine components.

The front of the structure is hinged on 38.1 mm (1.5 inch) square steel tubes welded to the front frame rails. The hood latches, strikers, cables and release handle are the type used on the 1994 Corvette. The hood release handle is mounted in the lower rear surface of the left front wheel well.

The striker pins had to be positioned on the rotating structure so that they would enter the stationary latches perpendicularly as the hood closed. Also the spacing between the striker pins and the latches had to be exactly the same in order for them to function properly.

The structure is fabricated from 19 mm (.75 inch) square steel tubing with 19 mm x 38.1 mm (.75 inch x 1.5 inch) steel struts and 19 mm (1.5 inch) square steel hinge columns all welded together. The hood structure has a check cable on each side to prevent over travel. The hood contains two gas assist springs that provide a high force to assist with opening the hood. The gas assist springs were developed in conjunction with Gary Weber at Arvin Meritor in Marion, South Carolina. Drawings of the hood structure and the vehicle frame along with the projected weight of the hood assembly (support structure and fiberglass panel) were sent to Gary. He used the information in a proprietary computer program and identified a production gas spring that he recommended for our application. The hood can be easily removed from the vehicle by disconnecting the two check cables, disconnecting the headlamp wiring connectors and removing the two hinge bolts.

Hood support structure with gas strut in the closed position.

Hood support hinge and gas strut in the open position

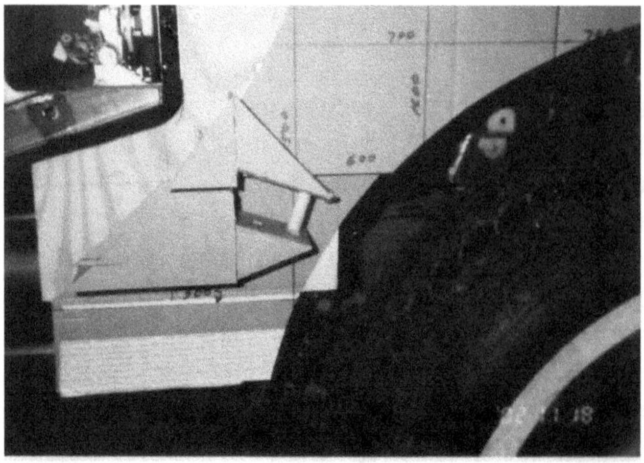

Rear deck alignment pin mocked up in plywood, pressed board and dowel rod shown in the closed position.

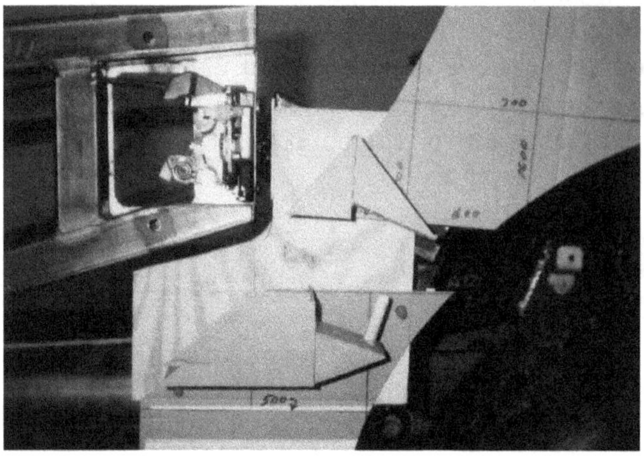

Rear deck alignment pin mock up shown in the open position.

The rear deck support structure is similar in construction to the hood support structure. The rear deck latch system is incorporated into pockets molded into the fiberglass rear deck panel and the fiberglass rocker panels. The latch strikers are attached to steel alignment brackets located on the under side of the fiberglass rear deck panel. The latches are attached to steel brackets located on the rear surface of the fiberglass rocker panels containing stainless steel alignment pins and rubber bumpers. The stainless steel pins enter holes in the rear deck alignment brackets to ensure the rear deck sides are flush with the rocker panels when the rear deck is closed. The rubber bumpers ensure the design gap is maintained between rear deck and rocker panels and they ensure the latches are in tension with the strikers. The rear deck support structure contains a steel sheet metal partition to separate the fuel tank area from the passenger compartment. The partition closes against an elastomer edge seal. The rear deck contains a chain and draw bar coil spring check system that is intended to cushion the stop as the rear deck reaches the full open position. The springs are wrapped with foam and leather to minimize rattles.

Mock up of rear deck latch pocket.

Completed rear deck alignment pin and rubber bumper- both with stainless steel hardware.

Rear deck tether system features drawbar springs to absorb the shock as the rear deck is opened.

Fitting Body Panels

The hood, rear deck, passenger side door and rocker panel were received at the FedEx depot in Romulus on 20 February 2004. We transported them to the workshop on a trailer. Fitting the body panels required that they be accurately temporarily held in relationship to each other and the under lying support structure. This necessitated that measurements be taken between the body panels and the steel structure and the shop floor. A centerline was scribed on the shop floor by dropping perpendiculars from the front and rear cross members of the vehicle frame. Centerlines were scribed on the hood panel and the rear deck panel using a pencil and a steel flexible straight edge (100 foot steel measuring tape). The centerlines on the hood and rear deck were carefully aligned with the centerline on the shop floor using a plum bob.

Plum bob on hood centerline aligns with vehicle centerline on garage floor.

The fore & aft locations and heights were established by aligning the "O-O" holes in the rocker panel mounting flanges with the "O-O" holes in the rocker panel mounting rails. The rocker panels were temporarily attached to the mounting rails using # 10 hex head sheet metal screws. Temporary metal and wood brackets were attached to the tail lamp surfaces and the headrest surfaces on the rear deck. The tail lamp surfaces would later be cut out to make room for the actual tail lamps. The holes in the headrest surfaces would later be covered with the upholstered headrest pads. The wood and metal brackets were attached to the steel support structure. Threaded studs and nuts permitted the rear deck to be moved and then precisely held with respect to the support structure and the adjacent rocker panels.

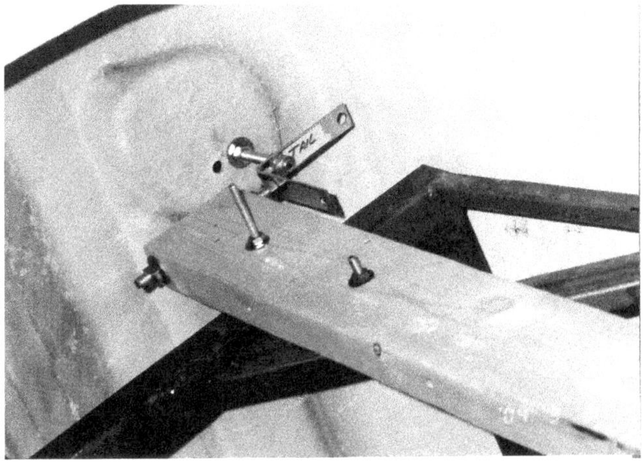

Temporary positioning system used to hold rear deck panel for bonding to the support structure.

A concern was identified during the early phase of fitting the rear deck. It was found that the lower portion of the fiberglass rear deck interfered with the steel pivot ends of the rear deck support structure. This was caused by a failure to cut sections at the pivot locations during the design phase. A centerline section indicated adequate clearance to the pivots however the lower surface curves forward at its outer ends and the actual condition was interference. Redesigning and revising the lower pivot points by moving them forward in the vehicle resolved the concern. This required cutting and welding the ends of both the rear deck support structure pivot columns and the pivot columns on the vehicle frame.

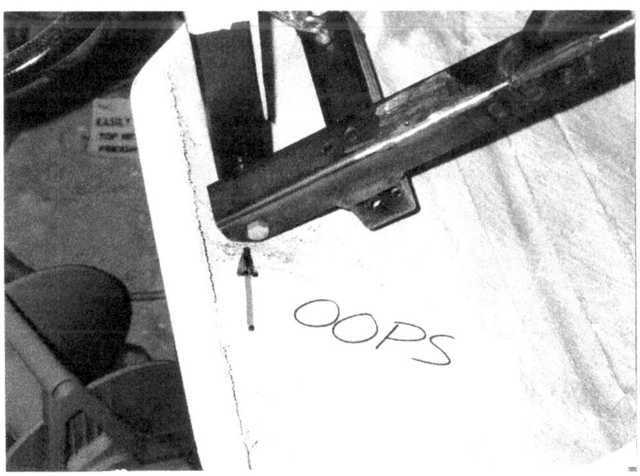

Interference identified early in the panel fitting process.

Metal brackets were temporarily attached to the headlamp surfaces in the fiberglass hood and to the steel headlamp mounting brackets in the steel hood support structure. The headlamp surfaces would later be cut out of the fiberglass hood to make room for the actual headlamps. Threaded studs and nuts permitted the hood to be adjusted in two planes (fore & aft and up & down) and then precisely held in the desired position.

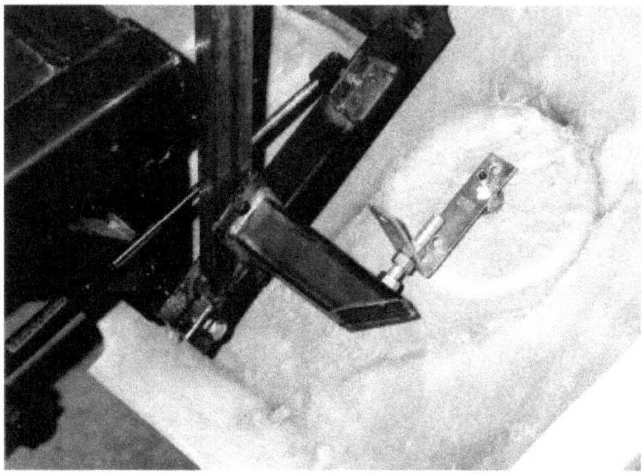

Temporary adjustable bracket system for bonding the hood panel to the support structure.

The trailing edge of the hood was positioned in its design location by placing it against a wood and steel bracket attached to the steel support structure at the vehicle centerline. Out board studs and nuts were used to precisely control the heights of the two rear corners of the hood. The two studs were threaded into tapped holes in the hood support structure.

The front wheel opening "tips" of the rocker panels were positioned and temporarily held in place by wood and metal brackets attached to the engine. Screws and nuts permitted adjustment to obtain and hold the precise side-to-side location of the tips.

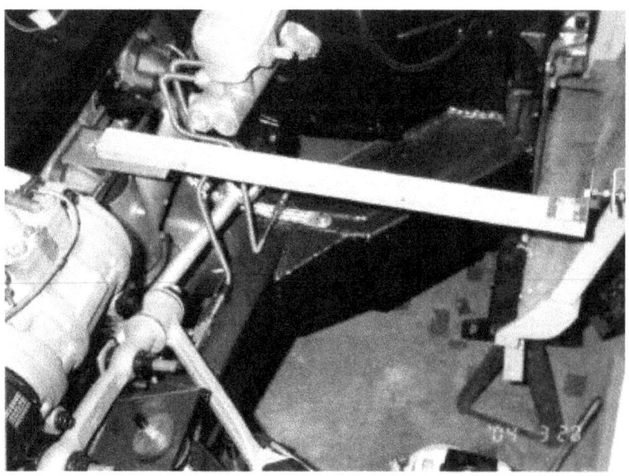

Adjustable fixture system to position and hold the rocker panel tips.

The fiberglass door panels were temporarily attached to the steel door support structure using a system of 90 degree steel brackets and studs and nuts. This system permitted adjustment and precise holding of the fiberglass door panels with respect to the door support structures.

Door panel position and hold system.

After the body panels were adjusted to achieve the intended alignment "target locations" were marked with a felt tip pin on the under side of the body panels for adding steel mounting brackets. The panels were then removed from their temporary mounting positions. Fiberglass pads were added to the under side of the fiberglass body panels at each of the target locations. The fiberglass pads were needed to minimize the possibility of "bleed thru" where the steel brackets would be bonded to the panels. Bleed thru can occur when high, concentrated forces are placed on the underside of fiberglass panels resulting in cosmetic marks showing on the top surface of the panel. The fiberglass pads were approximately 152 mm (6 inches) square and were "laid up" as follows: a coating of vinyl ester resin to the target area, 1 layer of fiberglass mat, 1 layer of core mat 3 mm (.12 inches) thick, 1 layer of fiberglass mat, 1 layer of fiberglass fabric (6 oz/yd^2) with a coat of vinyl ester resin between each layer and on top of the fiberglass fabric.

Typical body panel mounting bracket and pad.

The body panels were then reattached to their support structures using the temporary systems described above. Steel, 90 degree brackets, 32 mm (1.25 inches) wide x 38 mm (1.5 inches) x 38 mm (1.5 inches) x 1.6 mm (.06 inches) thick were modified by grinding the galvanized coating from one of the outer faces. Each bracket was carefully fitted to its location by drilling a 6.3 mm (.25 inch) hole in its galvanized face. The location of the hole had to result in the bare face being approximately 1.5 mm (.06 inches) from the fiberglass pad when the bracket was bolted to the steel support structure. The steel support structure had 6.3 mm (1.25 inch) holes drilled at the target locations to receive the brackets. M6 hex head cap screws were installed in the steel support structure. The length of the exposed thread was such that a hex nut could be installed to hold the cap screw in place and there would still be enough thread for the steel mounting bracket and a hex nut to hold the bracket in place.

A layer of methacrylate adhesive was applied to the bare face of the steel bracket (the literature and our own testing verified that the methacrylate bonds well to bare steel and poorly to galvanized steel). The bracket was placed over the exposed thread on the cap screw and a hex nut was tightened to hold the bracket in place while the adhesive cured. This process was repeated at each of the four corners on the hood and at four corners on the rear deck. The temporary mounting systems were removed and the hood and rear deck body fits were verified. The fits were judged to be satisfactory so the remainder of the steel brackets were added to the hood and rear deck to fully secure them in place.

Three, steel sheet metal "ribs" were added to each door structure. The ribs served as mounting surfaces for the steel 90-degree brackets. The fiberglass door panels were bonded in place using the same process that was used on the hood and rear deck. The door panels required extensive reworking of their edges and upper surfaces to achieve the desired fit to the adjacent hood, rocker panel and rear deck panels.

Three vertical steel ribs support each fiberglass door panel.

The rocker panels were bolted securely to the rocker panel mounting rails using M6 cap screws. The front regions of the rocker panels were attached to sheet metal structures that serve as inner fenders. In addition, the rocker panels were attached to the door pillars in front with sheet metal ribs. They were also attached to the rear bulkhead with sheet metal ribs.

Steel ribs attach the rocker panels to the door pillars.

Door lower and rear surfaces are fabricated from sheet steel and attached to the support structure with pop rivets.

Steel panels cover the mounting brackets on the hood.

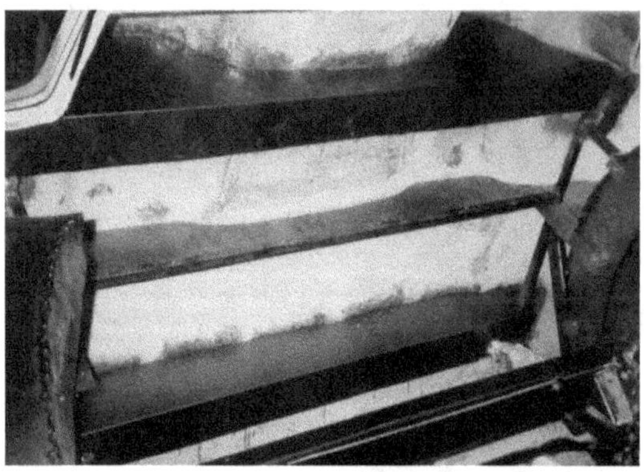

Rear deck with bracket covers and inner fenders in place.

Trial fitting the first two body panels received.

Inner Fenders

The inner fenders were developed on the vehicle and designed on the full size layouts. They are fabricated from sheet steel components welded together using continuous welds.

The rear inner fenders are curved in the side view and intersect and touch the curved inner surface of the fiberglass rear deck. The development of the line that intersects with the fiberglass surface was accomplished by building a wooden fixture that temporarily attached to the rear deck steel support structure. Measurements were made at intervals between the fixture and the fiberglass. The dimensions were transferred to corrugated paperboard models that were then fitted to the rear deck supports to verify the fit. After adjustments were made to the corrugated paperboard to achieve the desired fit the pattern was transferred to steel sheet. The rear inner fenders are attached to the steel support structure using screws and brackets. Their outer edges are adhesively bonded to the fiberglass surface.

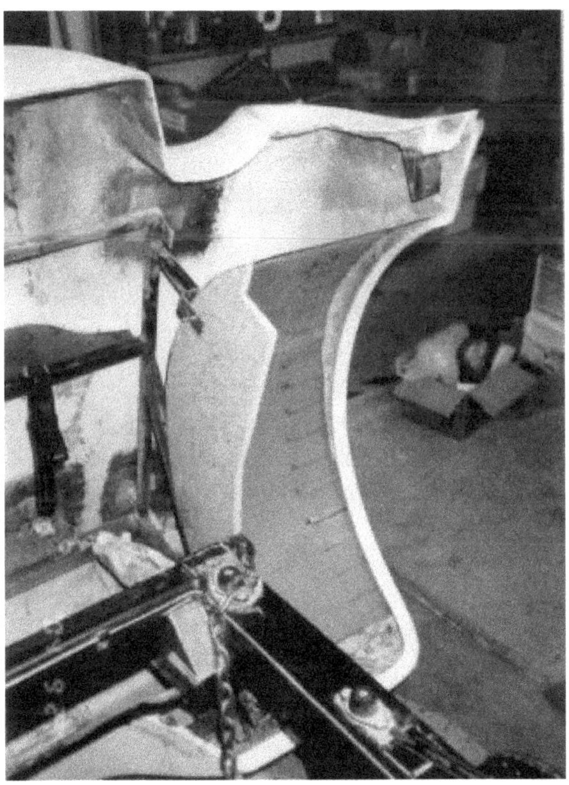

Developing the rear inner fender using OSB plywood and corrugated paperboard

The rear inner fenders are made of three sheet metal components welded together.

The front upper inner fenders are curved in the side view and were developed using the same technique as the rear. They are attached to the hood steel support structure using rivets and their edges are adhesively bonded to the inner surface of the fiberglass hood panel.

The front lower inner fenders were developed as a series of flat planes that intersect and touch the fiberglass rocker panels. The panels are attached to the frame using screws and rivets. They are attached to the fiberglass rocker panels using 90 degree steel brackets that are adhesively bonded to the inner surface of the rocker panels. The lower inner fenders are removable for access for service if necessary.

Front upper inner fenders are attached to the support structure using pop rivets.

Front lower inner fenders are attached to the frame with screws to permit access for service.

Fuel Filler Door

The initial design assumptions were to access the fuel filler neck by opening the rear deck like on one of the early Lamborghini vehicles. However after discussions with the team members it was decided to add a fuel filler door to the rear deck. This was because it is necessary to open both vehicle doors before the rear deck can be opened. It was agreed that it would be too bothersome to fill the fuel tank if the rear deck had to be opened. The aluminum door selected is the type used on the Mustang Bullitt. A production door assembly was modified by polishing it and laser etching the team logo on its top surface. The door cross sections are essentially a perfect match to the surface of the rear deck. A full size design layout was prepared to determine the optimum location of the door opening with respect to the fuel tank filler neck. The outline of a filling station fuel filler nozzle was traced on to a sheet of corrugated paperboard and cut out. The paperboard nozzle was used to identify the target "pierce point" in the rear deck. A mock up of the rear deck surface was fabricated in corrugated paperboard and placed in design position on the fuel tank. The opening diameter and location was then evaluated and judged to be excellent.

Verifying the pierce point for the fuel filler door was correct before machining the hole in the fiberglass rear deck.

Before the hole and counter bore were routed the under side of the rear deck had to be thickened to provide sufficient material for mounting the door. The target area was thickened by adhesively bonding a 250 mm (10 inch) diameter x 5 mm (.19 inch) thick pre molded vinyl ester/fiberglass "plate" to the under side of the deck. It was necessary to design and fabricate a fixture to route the hole and counter bore in the fiberglass rear deck. The fixture was fabricated out of plywood and temporarily attached to the rear deck using screws. The hole and counter bore were routed then the fixture was removed. Holes for the attaching screws were then drilled in the flange formed by the counter bore.

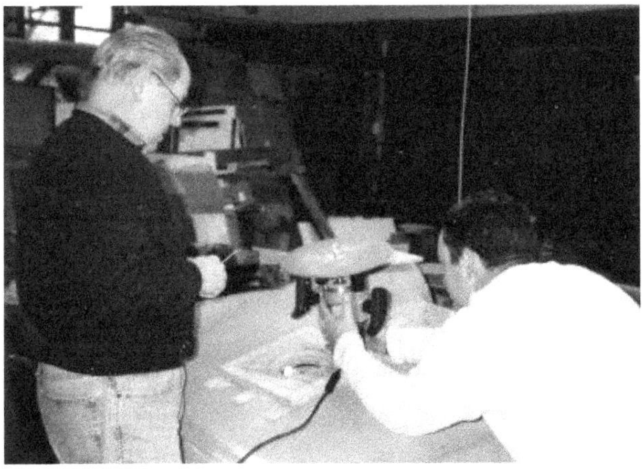

Machining the opening in the rear deck for the fuel filler door using a jig and router.

Grill

The grill was styled by Craig Sandvig. It was designed on a full size layout and CAD details were drawn. The grill follows the contour of the opening in the fiberglass hood in the side view. Each bar of the grill is located 1 mm (.04 inches) rearward of the one below it. The grill was mocked up using black tape 3.1 mm (.125 inch) wide over a sheet of pressed paperboard that had been painted silver. The first mock up was reviewed in the vehicle and the spacing between the bars was judged to be too large.

A second paperboard mock up was fabricated showing the bars closer together and reviewed in the vehicle. The new bar spacing was judged to be excellent. A mock up was fabricated using aluminum bars held in place with all threaded rods and screws. The "fabrication" holes in the mock up could be seen and were visually objectionable. The final grill is fabricated from aluminum flat stock components welded together. A fixture was fabricated to hold the bars in their design position during the welding process. The horizontal surfaces of the grill bars are painted flat black. The grill is attached to the steel hood support structure on the top and to the fiberglass hood on the bottom. Two steel brackets are adhesively bonded to the fiberglass hood panel for the lower mounts. Nuts and bolts are use to fasten it at its four corners.

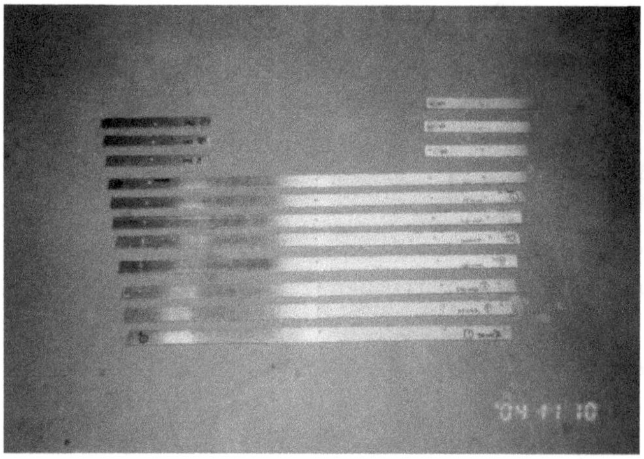

Aluminum components for the first grill that was fabricated.

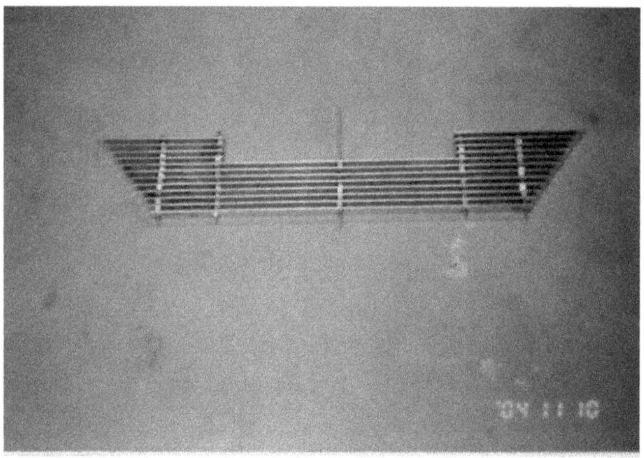

First grill as a welded assembly.

First grill was unacceptable because the fabrication holes were visible.

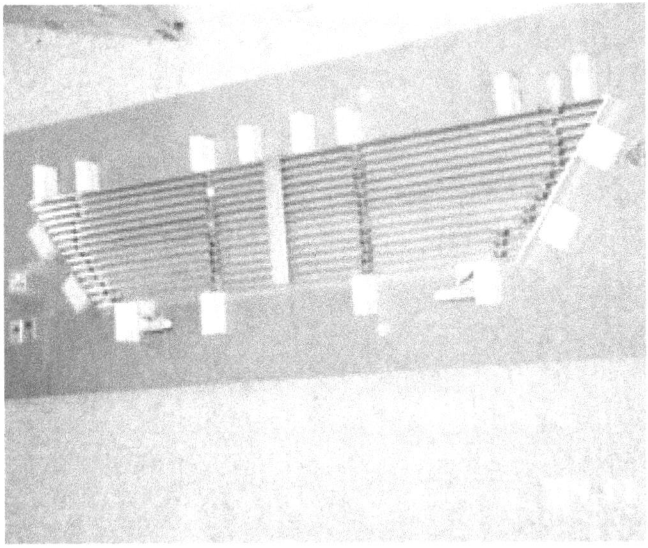

Final aluminum grill in the weld fixture.

Paint

The upholstery, steering column and instrument panel were removed from the vehicle in preparation for paint. A temporary panel was installed on the transmission tunnel to support the ignition ON/OFF switch and the START switch. The steering column and steering wheel from the donor vehicle was installed in the vehicle. Temporary wiring was installed for the ignition circuit and brake lights. The vehicle was driven to Biggs Auto Renovation in Westland, Michigan on 6 September 2005 (approximately 5 miles in LOW gear all the way because we had removed the speedometer electronic module for service and thus inadvertently broke the communication link between the PCM and the electronically controlled transmission). Swane Clayton began the painting process on 7 September 2005 and completed his work on 17 October 2005.

The paint color is GM Torch Red.

Drivable vehicle with rocker panel in primer.

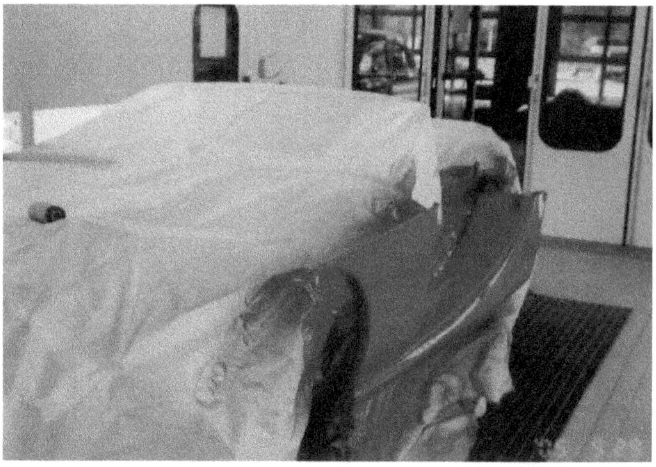

Rocker panel with color coat applied.

Swane Clayton polishing the paint on the rear deck.

Final assembly after paint.

10

Cockpit

The instrument panel was developed using full size layouts, reduced scale layouts, wood and corrugated board mock-ups and CAD tools. The first step in developing the instrument panel was to establish the plane of the gauges and the field of view thru the steering wheel. The gauge plane was located 260 mm (10.24 inches) forward of the front plane of the steering wheel and perpendicular to the centerline of the steering column. The field of view was established on the layout and verified by fitting a corrugated paperboard "plane" to the vehicle.

Mock up to verify the field of view of the gauges thru the steering wheel.

The gauges to be used were selected from the catalog of Classic Instruments. The gauges were drawn in on the layout using the published dimensions. Paper cut out drawings of the gauges were added to the corrugated paperboard plane to verify readability from the driver's seat.

A mock up of the rough architecture of the instrument panel was fabricated from 12 mm (.50 inch) plywood, and 50 mm x 50 mm (2 inch x 2 inch) pine boards. The rough instrument panel mock up was fitted with a piece of 3 mm (.12 inch) particleboard with holes drilled to accept 3 dimensional mock-ups of the gauges. The gauges were mocked up using corrugated paperboard and hot melt adhesive. The gauge mock-ups were fitted to the cluster mock up to verify clearances to the steering column mounting brace.

First mock up the instrument panel to verify the location of the major planes.

Copies of the paper layouts that identified the locations of the "hard points" were given to Craig Sandvig. He would use the layouts for developing the surfaces of the instrument panel. Craig developed the overall "theme" of the panel and some of the details.

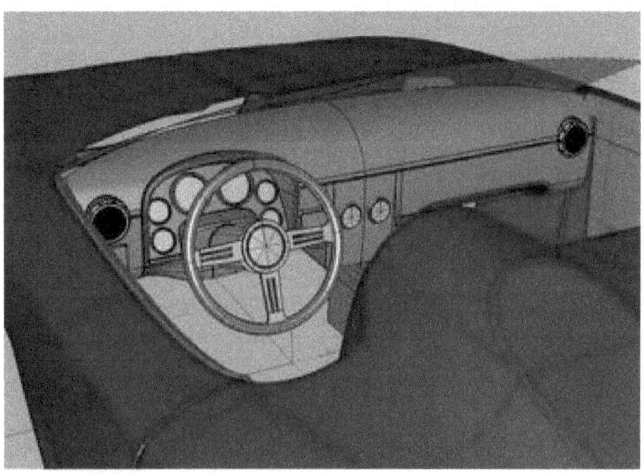

Overall instrument panel theme was established with a CAD tool.

Craig transferred the information to Larry Conger who then developed the final surfaces of the instrument panel. Larry designed the panel on his computer taking into consideration the inside surface of the windshield, the door trim panels, the windshield support structure "A Pillars" and the transmission tunnel. Larry also "stepped the surface" of the upper portion of the instrument panel 1 mm (.04 inch) to accept the leather trim. After a review of Larry's CAD drawing the team recommended some revisions. Larry made the revisions and then sent the CAD file to Chris Mason in North Carolina on 27 May 2004. The molded fiberglass instrument panel was completed on 26 July 2004 and shipped via UPS to our workshop.

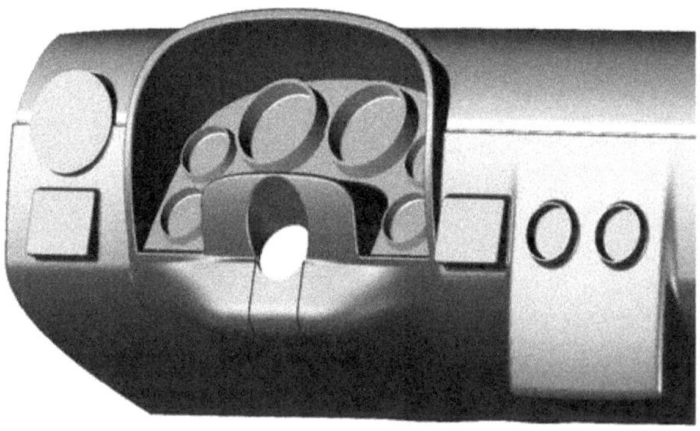

CAD model of the instrument panel surface.

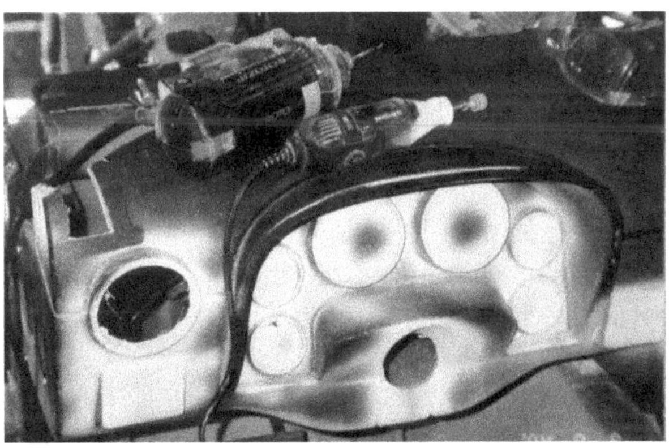

Holes were cut for the gauges, switches and registers using rotary cutting tools.

A trundle fixture allowed the panel to be supported and rotated for bench work.

Upholstery Substrates

The door upholstery trim panel substrates were fabricated from 5 mm (.20 inches) Lauan, 12 mm (.50 inch) plywood, vinyl ester resin / fiberglass cloth and sheet metal.

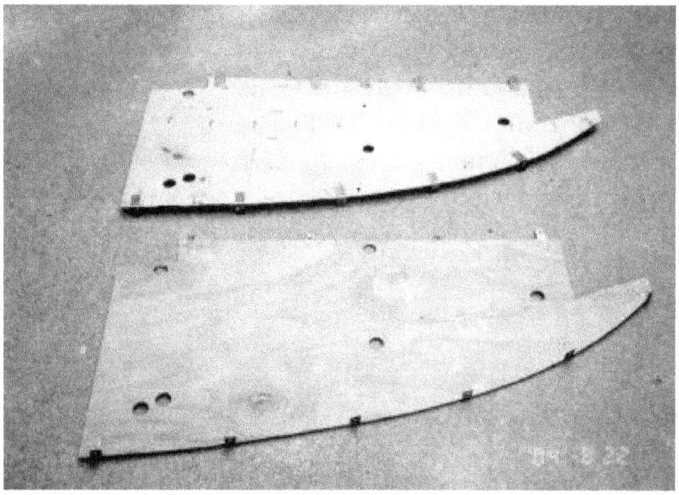

Door trim panel upholstery substrates were fabricated using 5 mm Lauan plywood.

The armrests were the first substrates to be designed. They were designed on a full size layout. They were developed by setting in the vehicle on a mock up seat and trying several shapes, sizes and locations. The armrests were intended to provide a resting perch for the occupants' arms, to provide ready access to the door handle and to provide a door pull for closing the door. After the basic size and location were developed, Craig Sandvig prepared sketches of the styling features for the armrests. Each armrest was fabricated beginning with a 5 mm (.20 inch) Lauan back plate. Two top plates were cut from 12 mm (.50 inch) plywood. The lower of the two top plates was bonded to the back plate using hot melt adhesive and wood screws.

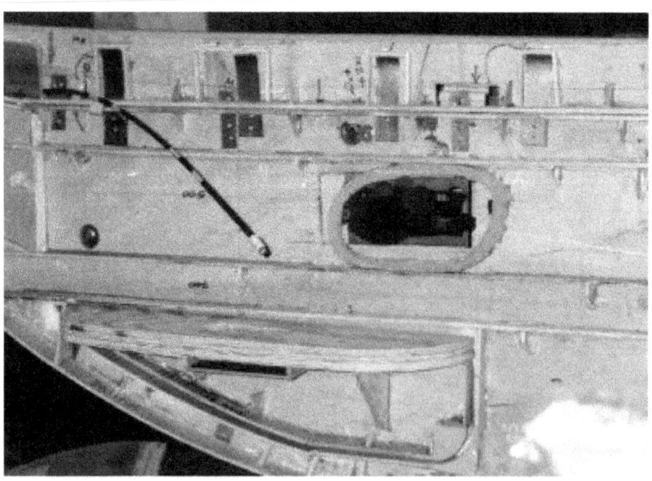

Door trim panel upper and lower substrates with arm rest in place.

Steel brackets tie the arm rests to the door structure.

A pocket was formed under the lower top plate to receive a steel-mounting bracket. A lower edge was fabricated using 5 mm (.20 inch) Lauan with saw kerfs partially cut to permit bending. The lower edges were attached the back plate using hot melt adhesive. Three 1/4- 20-cleat nuts were attached to the back plate and held in place with hot melt adhesive. The back plate was "loaded" with slabs of polystyrene foam held in place with hot melt adhesive. The foam was shaped using a Sureform tool. The foam was smoothed using dry wall compound.

Latex primer was applied to the foam. Two layers of vinyl ester resin and fiberglass cloth were applied and sanded to complete the armrest substrates.

Each armrest is attached to its respective door trim panel with three flat head 1/4 – 20 machine screws driven into the cleat nuts. In addition each armrest is supported with a steel bracket that projects from the door steel structure. Each steel bracket has two 1/4-20 J nuts that receive screws passing thru two counter sunk holes in the plywood lower plate. A door pull trim cup is attached to the lower plate using two wood screws.

The door trim panel substrates were fabricated beginning with a 5 mm (.20 inch) Lauan back plate. Each back plate is flat and covers the lower portion of the door structure. Each is held in place with steel tabs, pop rivets and sheet metal screws. Each upper door trim panel is curved to match the inner flange of the fiberglass door trim panel. Each upper trim panel substrate was cut from 5 mm (.20 inch) Lauan. It was then "notched" with partial saw kerfs to permit it to easily bend. A 1.2 mm (.048 inch) thick chipboard spacer was temporarily attached to the fiberglass door flange using double face tape. The spacer represented the

A "pocket and flange" was fabricated in each of the lower trim panels to receive the door handle trim bezels. The pockets were fabricated from 5 mm (.20 inch) Lauan and 3 mm (.125) thick pressed board. They are attached to each back plate using hot melt adhesive.

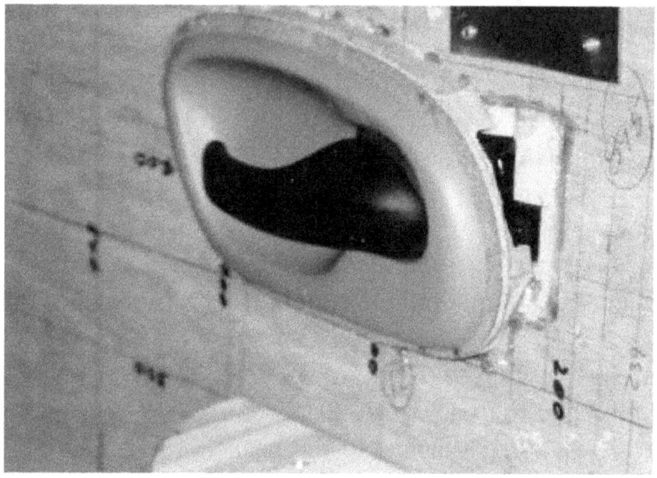

Pocket and flange for attaching the door handle trim bezel.

A flange was added to the lower edge of the lower trim panels using the partial kerfs technique described earlier. Flanges were added to the lower edge of the upper substrates using 5 mm (.20 inch) Lauan. Matching flanges were added to the upper edge of the lower trim panels using 5 mm (.20 inch) Lauan. The two flanges were off set by 2 mm (.08 inches) to provide room for the two layers of leather that would be added. The upper and lower substrates were then "loaded" with polystyrene foam and shaped and finished the same as the arm rests discussed above.

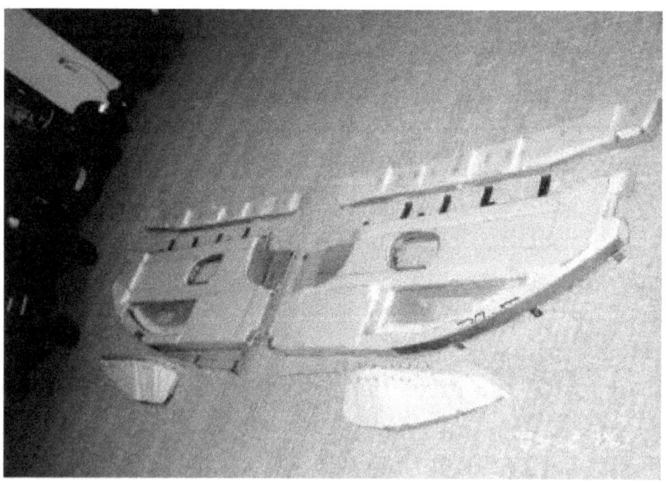

Substrates "loaded" with foam and ready for shaping and smoothing.

The substrates for the lower edge of the rear deck were fabricated using the “splash” technique. Sheet wax 1 mm (.04 inches) thick was applied to the areas to be covered. The sheet wax served as a spacer for the leather that would be applied to the substrate. Six layers of vinyl ester resin and fiberglass cloth were applied to the sheet wax. After the resin cured the substrates were removed and trimmed to size. End “transition” pieces were formed using 12 mm (.50 inch) plywood, 5 mm (.20 inch) Lauan and polystyrene foam. They are attached to the vinyl ester resin substrates using methacrylate adhesive. The complete substrates are attached to the rear deck fiberglass panels using 1/4- 20 hex head screws and cleat nuts at their outer ends. They are tucked under the headrest substrates at their inner ends.

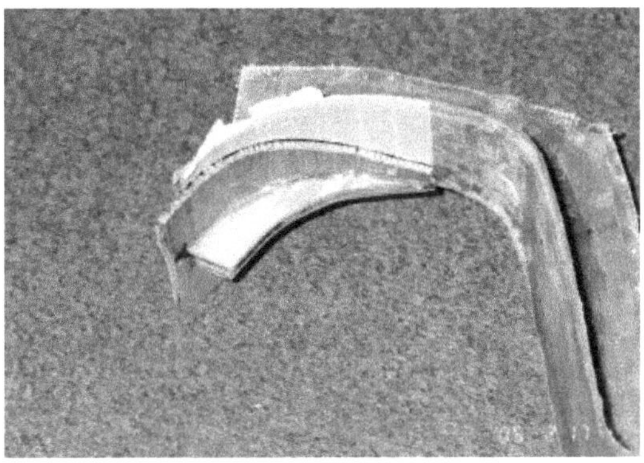

The substrate for the lower edge of the rear deck was fabricated using the “splash” method.

The upholstery substrate that covers the instrument panel top was fabricated using the splash technique described above, i.e. sheet wax, 5 layers of vinyl ester resin and fiberglass then smoothed.

The console was developed on full size layouts and using mock ups in the vehicle. The first mock up was fabricated using corrugated paperboard. After reviewing the first mock up it was decided to revise the cross section of the console. A second paperboard mock was fabricated and it was agreed that it would be the final design.

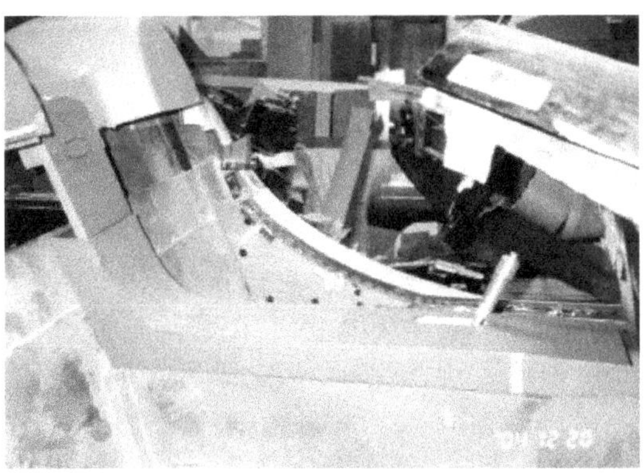

Final mock up of the console fabricated using corrugated paper board.

Fitting the console to the transmission shifter required a careful, detailed design. The console lower surface had to be above the shifter mechanism, the opening had to be precisely located fore and aft and side-to-side and the upper surface of the console had to be as close as possible to the shifter mechanism to maximize the space for the leather shifter boot. A partial mock up of the console in the immediate area of the shifter lever was fabricated using 12 mm (.50 inch) plywood. The partial mock up verified the functional conditions were met. The precise location of the opening was then determined by measuring the distance from the opening to the intersection of the rear bulkhead and the transmission tunnel top. The console was then fabricated from 12mm (.50 inch) plywood. The curved face at the upper rear of the console was accomplished using the partial kerfs technique. The partial kerfs were filled with polyurethane adhesive after the top panel was attached to the side panels using dry wall screws. The console substrate was then smoothed using polyester body filler.

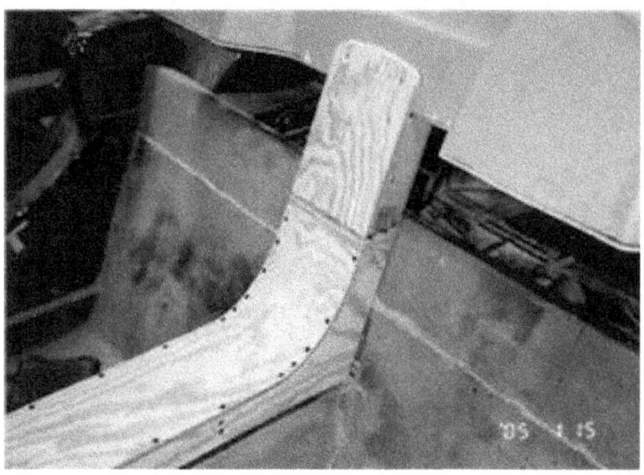

Console substrate fabricated in plywood.

The upholstery substrates for the doorsill trim covers were fabricated from sheet metal. The curved panels were formed and clamped in a weld fixture. The flat side panels were then skip welded to the curved panels to minimize warpage.

The substrates for the moldings that trim the door striker covers were fabricated from sheet metal using a technique similar to the one used for the doorsill trim covers.

The windshield “A Pillar” trim cover substrates were fabricated in sheet metal and skip welded together. They slip over the windshield supports and are held in place with the outside rear view mirror mounting bolts at the top and with M6 cap screws into the lower windshield support structure. The attaching fasteners are hidden by the rear view mirror trim bezels and the lower windshield trim cover.

Mock up used to determine the exact location of the shifter lever opening with respect to the tunnel.

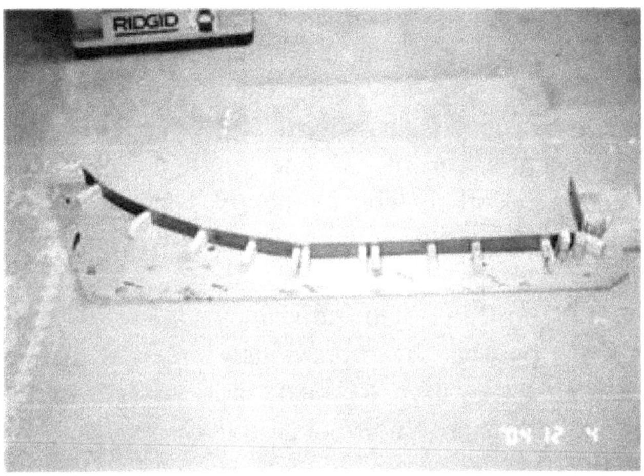

Door sill trim cover substrate in weld fixture.

Doorsill substrates were skip welded to minimize warpage.

"A Pillar" trim substrates were fabricated in sheet steel and welded.

Upholstery

The first phase of the upholstery application (leather, less carpet) began on 18 July 2005. Mark Bowden at Mobil Trim in Garden City, Michigan developed the patterns for covering each component in vinyl or scrap leather before he cut, sewed and fitted the finish leather. Upholstery foam at various thicknesses was used between the substrate and the leather. The leather is sewn with polyester thread. Polyester thread was selected because it is more resistant to degradation due to the sun versus nylon thread. The leather is attached to the substrates with adhesives and or stainless steel staples. The team logo had been previously embroidered on the piece of leather that was used to cover the console “waterfall”. American Silkscreen and Embroidery did the embroidery. Mark completed the first phase on 26 August 2005.

The vehicle was then returned to the upholstery shop on 2 November 2005 for the second phase of fitting the carpet and the trim covers for the windshield “A Pillars”. Mark Bowden completed the carpet portion of the upholstery project on 18 November 2005.

Mark Bowden discusses the first phase of upholstery with Chuck Carlson.

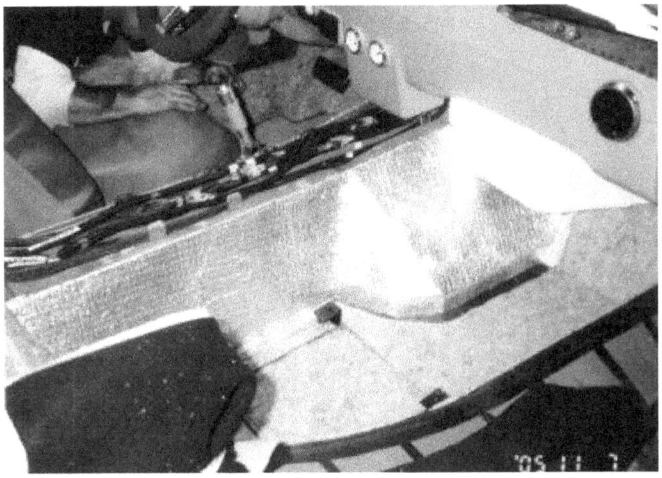

Heat shielding carpet padding was applied to each side of the tunnel prior to carpeting.

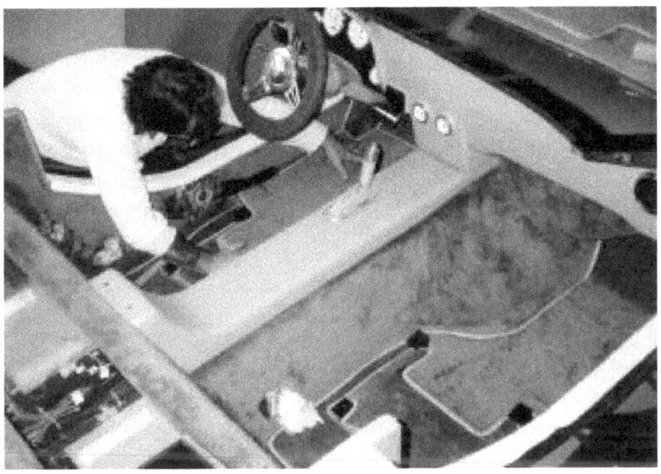

Carpeting was the final upholstery task.

Seats

Several seat alternatives were evaluated for appearance, size and seating comfort. The Cerullo Performance Seating model SC seat was selected for the vehicle. It was evaluated initially with an unupholstered sample seat sent by the Cerullo people. The evaluation included placing the seat in the vehicle and confirming it would fit between the transmission tunnel and the frame rails. The SC model is one of the few aftermarket seats fabricated 495 mm (19.5 inches) wide.

Most seats available are wider and would not package in the vehicle. The team judged the sample seat to be comfortable to set in and its appearance was well liked. Adapter brackets were fabricated to mount the Cerullo seats at the desired angle of 8.5 degrees from the horizontal and to position the seats as close as possible to the floor in order to achieve the desired H point. Two leather hides and four yards of vinyl in Light Oak color were shipped to Cerullo in Pomona, California for fabricating the seats for the vehicle.

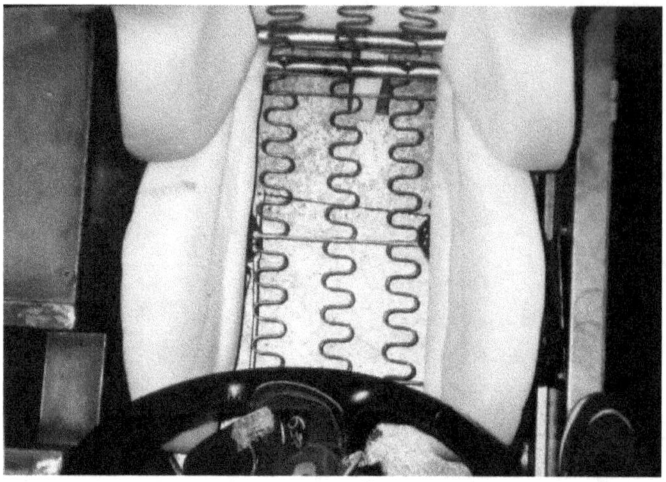

An unupholstered seat is trial fitted to verify clearance exists to the tunnel and frame rail.

Steel mounting brackets were used to tilt the seats 8.5 degrees.

Seat Belts and Shoulder Harnesses

The lengths for the seat belts and shoulder harnesses were developed in the vehicle using the final seats. The attaching points for the seat belts had been determined during the design of the frame and fasteners were welded in place during the fabrication of the frame and transmission tunnel. The attaching points for the shoulder harnesses were established using the final seats. The color for the seat belt webbing was carefully selected by the team from several samples provided by the supplier, Ssnake-Oyl Products.

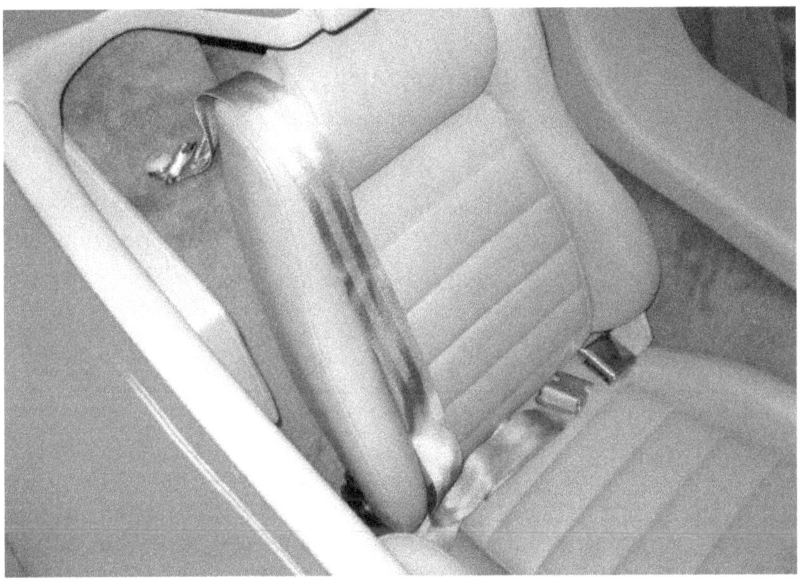

The seat belts and shoulder harnesses were developed using the final seats.

Transmission Shifter Knob and Boot

The transmission shifter knob was styled by Craig Sandvig. The components were designed on a full size layout and CAD details were prepared. The details were sent to several firms to quote. A.S.A.P. Machine Co. of Canton, Michigan was selected to machine the parts in aluminum. The knob components are adhesively bonded to an aluminum tube. The aluminum tube slides onto the Corvette shifter lever and is held in place with a setscrew. John Ledbetter at Laird's Glass & Upholstery wrapped the conical part of the knob in leather. John also provided several helpful suggestions related to the upholstery substrates.

The shifter lever is trimmed with a leather boot that is secured to the console top using a machined aluminum bezel supplied by Lokar.

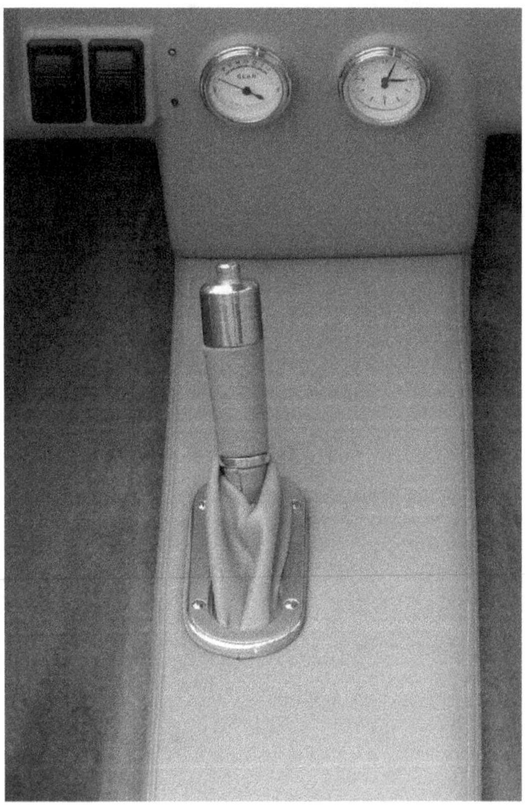

Shifter knob is machined in aluminum and wrapped with leather.

11

Components and Subsystems

Air Conditioning and Heating

Even though the vehicle is open it was decided to fit a heater and air conditioner in order to make it more comfortable for the occupants when driving in cold or hot weather. The major components of the heater and A/C system are recycled from the donor vehicle. The evaporator and blower assembly are repackaged to fit the new under hood environment.

The heater and A/C control system is new. It features a three-position rocker switch. In the HEAT mode the red LED is on and the temperature door is in the max warm position. When the rocker switch is in the mid position the system is OFF. In the A/C position the blue LED is lit and the temperature door is in the max cool position. The blower is on high speed in both the heater and A/C mode.

Air Conditioner Duct

The instrument panel A/C duct was designed using full size layouts and full size mock ups. The duct has a nearly constant cross sectional area its entire length and it has a 5 mm (.20 inch) clearance to the gauges, windshield support structure and the steering column. The duct was fabricated using pressed fiberboard, corrugated paperboard, hot melt adhesive, polyurethane adhesive and covered with vinyl ester resin and fiberglass fabric. The duct has a compression seal to the heater and A/C center distribution chamber. The duct ends slip over the outboard instrument panel registers. The center flange is located and supported with steel brackets and threaded studs attached to the fiberglass instrument panel.

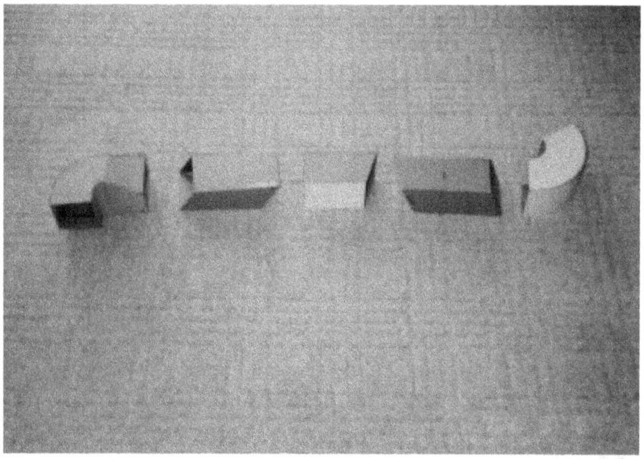

Components for instrument panel A/C duct built in corrugated paperboard.

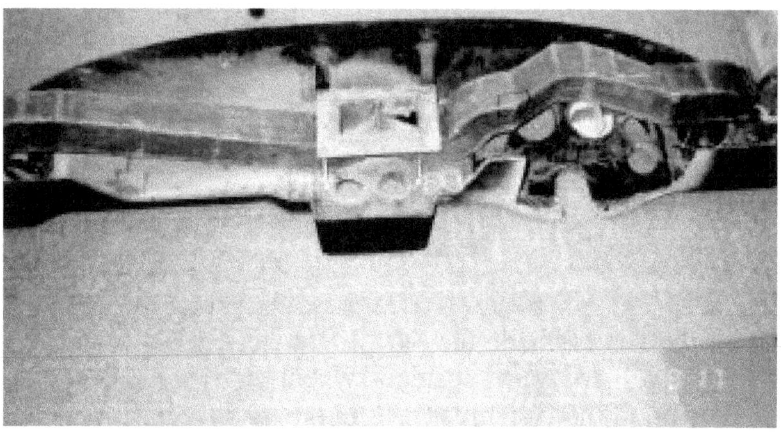

A/C duct covered with fiberglass cloth and vinyl ester resin to provide strength.

A/C and Heater Distribution Chamber

The heater and A/C center distribution chamber is a highly modified unit based upon the type used on the donor vehicle. The openings for the center floor outlet and the defroster outlet have been covered with sheet aluminum panels pop riveted to the plastic housing. New floor outlets have been fabricated using pressed fiberboard and polyurethane adhesive. They are pop riveted to the plastic housing. The floor outlets were designed to clear the transmission tunnel and direct appropriate size streams of air to the occupants' feet. The chamber is attached to the heater core case using a foam seal. The chamber is attached to the cross car beam using M6 hex head cap screws.

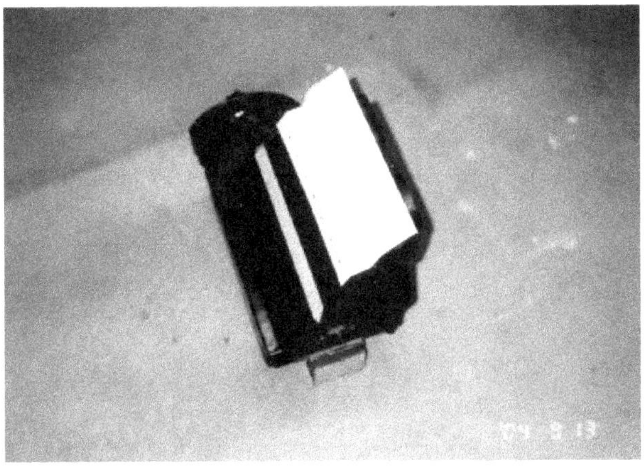

A/C distribution chamber was modified part from donor vehicle.

A/C Refrigerant High Pressure Cut Out Switch

The A/C refrigerant system is protected with a high pressure cut out switch. Should the refrigerant system pressure exceed 3100 kPa (450 psi) the electrical circuit to the A/C clutch will open and thus relieve the over pressure condition until repairs can be completed.

A/C Instrument Panel Registers

The A/C instrument panel registers are the type used on the 2005 Ford Freestyle. They were selected for their appearance and their functionality. They have shutoffs built in that permit the vehicle occupants to control the flow.

They are mounted to the instrument panel using 90 degree brackets attached to the register housings. Number 10 threaded rods and nuts hold the registers firmly in place.

Brake Rotors and Flex Lines

The brake rotors are cross-drilled cast iron units supplied by Power Stop division of TRW. The hubs of the rotors have been painted black. The brake fluid flexible lines have braided stainless steel covers and are supplied by Russell.

Cross-drilled brake rotors and stainless steel flex lines.

Brake and Fuel Lines

The brake and fuel lines were developed on the vehicle and designed on the full size layouts. Mock-ups of the lines were fabricated using 9.5 mm (.375 inch) diameter wood dowel rod and plywood corner braces and hot melt adhesive. Detail drawings were prepared of the tubes. In Line Tube located in Shelby Township, Michigan fabricated the tubes in stainless steel. The tubes are fitted with stainless steel coil spring stone shields. In Line Tube fabricated the tubes by scanning the wood mock ups rather than from the dimensions on the detail drawings.

Brake and fuel lines mocked up using wood dowel rods.

The brake and fuel lines are routed under the body thru three sided channels formed by the frame rails on the out side and special steps in the floor panels on the top and inside. The brake lines are protected from the heat radiated from the catalytic converter on the driver side by a steel shield fastened to the frame.

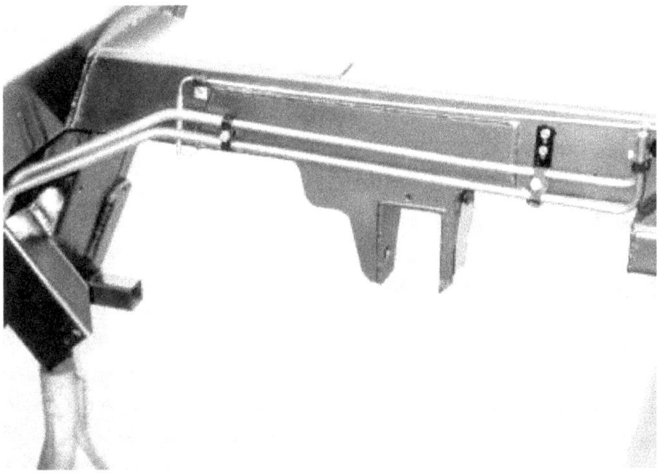

Brake and fuel lines fabricated in stainless steel.

Brake and Accelerator Pedal Pads

A new brake pedal pad steel substrate was developed to accept a new rubber pad like the type used on the 1968-1974 Corvette. A new hinge/spring mechanism was developed to accept an accelerator pedal pad from the 1968-1974 Corvette. The pads were selected because they feature bright stainless steel accent moldings.

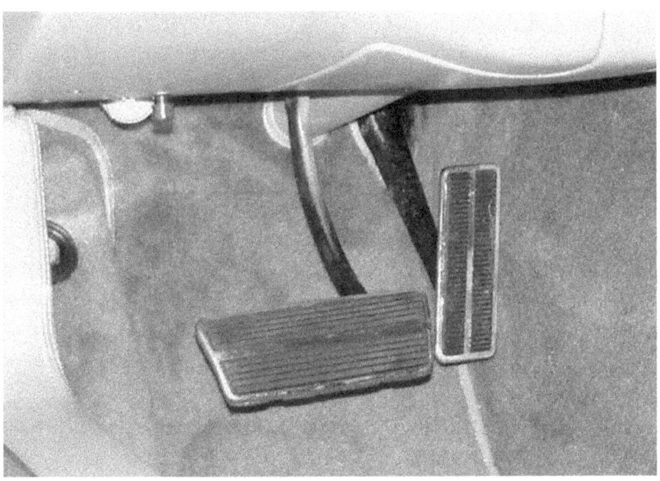

Stainless steel moldings trim the brake and accelerator pedal pads.

Electrical System

The electrical wiring system was designed on the computer and developed on a full size wood mock up of the electrical component mounting panel, the transmission tunnel top and the instrument panel. James Wilber helped with the design by physically identifying and marking a majority of the circuits on the donor vehicle wiring harness and by reviewing the 1994 Corvette Service Manual. One of the major elements of the electrical system is the Powertrain Control Module (PCM) or engine computer. The PCM is securely mounted in a steel compartment under the floor of the rear storage compartment and has wires running from it to the engine, transmission and to the rear electrical component-mounting panel. The PCM wiring harness is very specialized because of the unique terminals, connectors and wire lengths to fit the vehicle. In addition, the PCM software is specialized to make it compatible with the vehicle electrical system. Howell Engine Developments in Marine City, Michigan was selected to fabricate the PCM wiring harness and to modify the PCM software. James, Chuck and I traveled to Marine City and met with the people at Howell Engine twice before the design was finalized. The Howell Engine people were provided with the PCM from the donor vehicle, a circuit specification sheet, an engineering drawing of the wiring harness as well as a mock up fabricated from rope and wood blocks to simulate the electrical connectors to use for fabricating the wiring harness.

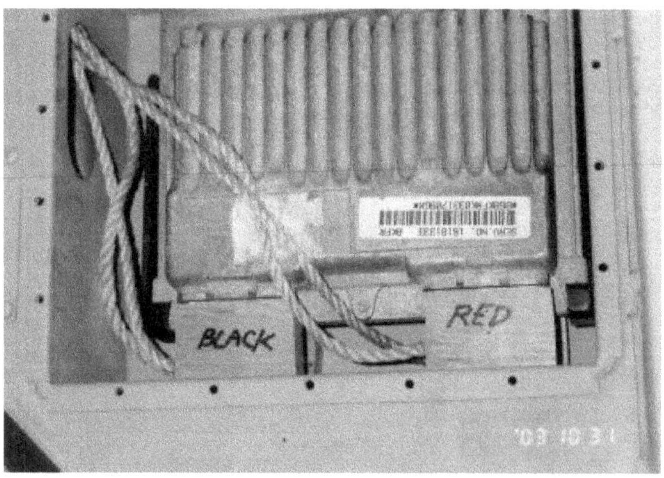

Developing wire routing and length using mock ups in wood and rope.

The catalogs published by the wiring component manufacturers and distributors proved to be very helpful for designing the new electrical system. The catalogs listed several alternative wiring terminal/connector families to choose from. They also listed the specialized crimping tools that were mandatory to achieve reliable crimps of the terminals to the wires.

The balance of the vehicle electrical system was designed around just four visible electrical switches on the instrument panel. The four instrument panel rocker switches were selected for their appearance and ergonomic feel. The START switch is located to the far left side of the instrument panel to minimize the risk of accidental actuation by the passenger. The start switch has a red actuator cover to further identify it as a special switch.

The ignition on/off switch has a green Light Emitting Diode (LED) mounted next to it that illuminates when the ignition is on. The climate control switch is a three-position rocker. When the switch is in the middle position the heater and A/C system is off and neither the red nor blue LED is illuminated. When the switch is in the down/on position the heater is on, the system blows hot air and the red LED is illuminated. When the switch is in the up/on position the system blows refrigerated air and the blue LED is illuminated. The wiring system to accomplish the simple on/off/on control is made up of relays and limit switches rather than an electronic module. It was a major challenge to design the control sub system because the polarity of the current feeding the temperature blend door must reverse in order for the door to move from the warm to cool or cool to warm positions.

The vehicle electrical system features 12 relays, 7 fuse blocks, 38 fuses, 5 terminal junction blocks, 240 electrical circuits, 8 LED's, 14 switches, 2 horns, a power point, a 1,000 cranking amp battery with both side and top terminals. Essentially every circuit has a self-adhesive number marker on each end of the wire. The electrical system is described on 41 computer drawn schematic sheets. Chuck Carlson drew the schematics using Microsoft Excel. The schematic sheets are included in the GDT Speedster Service Manual. In addition, spreadsheets were prepared that listed every circuit by number, the length of each wire, the wire color, the inlet terminal part number, the outlet terminal part number, the inlet connector part number and the outlet connector part number. This spread sheet was used to order the appropriate quantity of components to fabricate the system.

The head light dimmer switch is located in the steady rest for the left foot. It was carefully packaged so that the switch would not be accidentally actuated by the driver's left foot while in the normal rest position and yet would not be located so high that the driver's foot would contact the instrument panel during actuation.

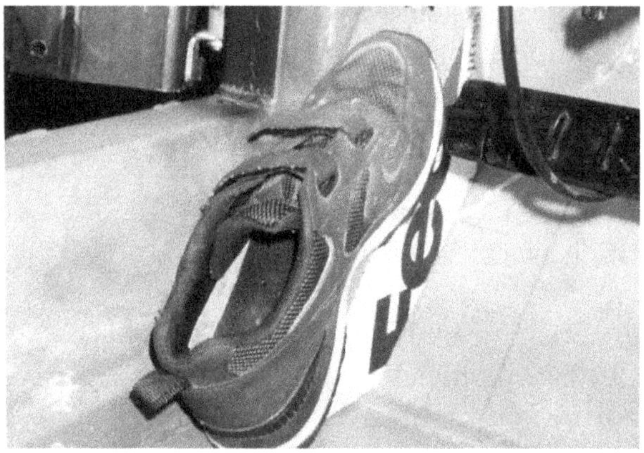

Developing location for headlamp dimmer switch.

The instrument panel gauges were selected from the catalog of Classic Instruments. The speedometer and tachometer were custom fabricated for the GDT Speedster with the team logo printed on their faces. The gauges were temporarily placed in a pressed board panel located in design position during the wire routing phase.

The vehicle features a "PRNDL" gauge located near the center of the instrument panel. The PRNDL gauge is electrically actuated from a signal from a potentiometer located on the gearshift lever mechanism. The design of the linkage to actuate the potentiometer and the packaging to clear the existing mounting pocket in the transmission tunnel was accomplished using full size layouts and full size metal mockups.

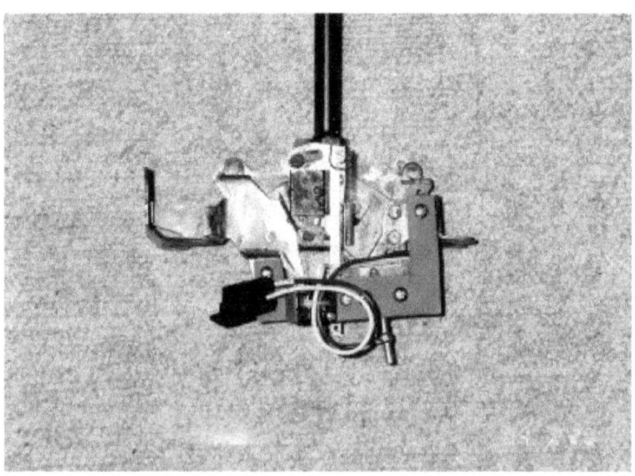

Potentiometer on transmission selector for inputting to the PRNDL gauge.

A series of live tests were started on 4 February 2004 to verify every electrical circuit functioned properly. The tests were conducted with the components mounted on the wooden mock up well before the wiring system was installed in the vehicle. The vehicle battery was used to supply the power to each of the circuits being tested. The tests identified a few concerns that were corrected before system was installed in the vehicle.

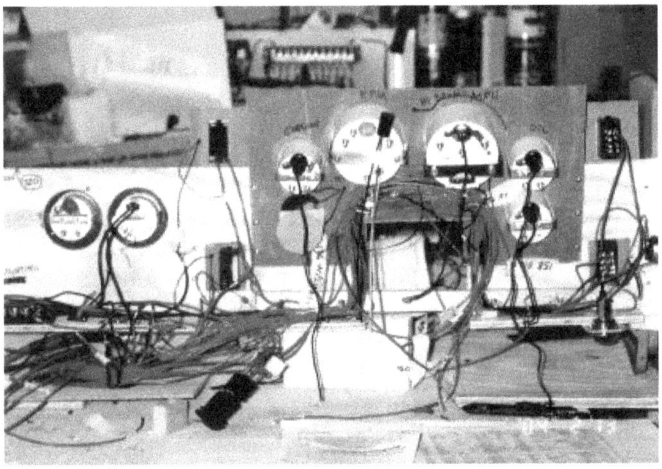

Developing wire length and routing on a wooden mock up.

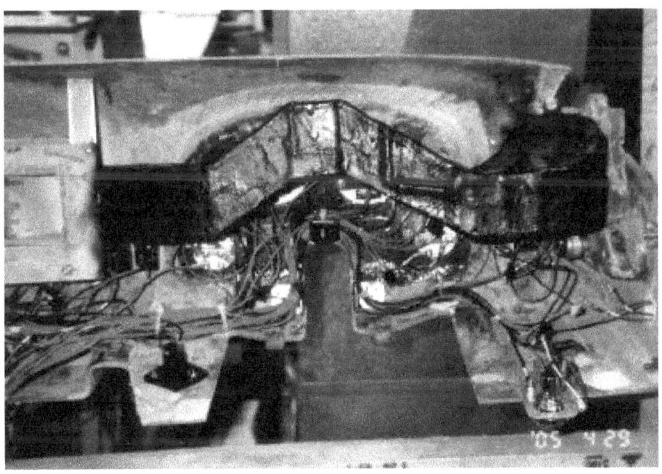

Completed wire routing in the actual instrument panel.

Electrical component mounting board is a major element of electrical system.

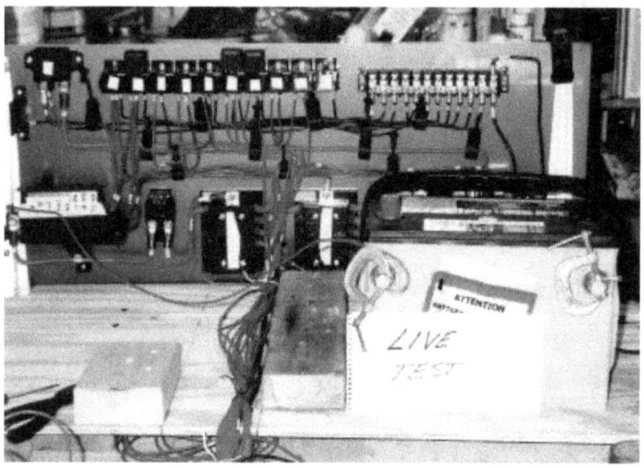

Live electrical test on bench to identify concerns where they were easy to fix.

Example page of spreadsheet used to tabulate every component in the electrical system.

WIRES/TERMINALS/CONNECTORS

Circuit	Description	Gauge	Length	Color	Input					Output				
			FEET		Terminal			Connector		Terminal			Connector	
					Type	Size	Part No	Part No.	Pin No.	Type	Size	Part No	Part No.	Pin No.
1	Battery (+) to starter motor Taylor kit 04800	2	8	Red	Battery Top			NA	NA	Eyelet	3/8	PIA kit	NA	NA
2	Battery (-) to frame Taylor kit 04800	2	3	Black	Battery Top		PIA kit	NA	NA	Eyelet	5/16	PIA kit	NA	NA
3	Battery to Maxi fuse block Lynx LCS 45, 07024	4	45 inch	Red	Battery Side		PIA cable	NA	NA	Eyelet	1/4 stud 4 Ga	36461	NA	NA
4	Maxi fuse block to 80 amp in-line fuse	8	1	Red	Ring	1/4 stud 8 Ga	33502	NA	NA	Ring	#10 screw 8 Ga	33501	on hand	in put
5	CIRCUIT DELETED													
6	Maxi fuse block fuse 4 to starter relay 3	10	1	Red	Butt	10 Ga	32960	NA	NA	Ring	(M5) #10 10 Ga	32702	NA hvyduty	NA hvy duty
7	CIRCUIT DELETED													
8	Maxi fuse block fuse 6 to alternator	8	9	Red	Butt Step	8 Ga to 10 Ga	38052	NA	NA	Ring	5/16 stud 8 Ga	33503	500057	Boot
9	CIRCUIT DELETED													
10	Maxi fuse block fuse 17 to ignition sw relay 2	10	2	Red	Butt	10 Ga	32960	NA	NA	AMP	10 Ga	180351-2	75280	30
11	Maxi fuse block fuse 16 to HIR&S relay 1	12	2	Red	Butt	12 Ga	32960	NA	NA	Amp	12 Ga	180351-2	75280	87
12	Cylinder block to frame Lynx LCS 14E, 04076	4	14 inch	Metal	Eyelet		PIA cable	NA	NA	Eyelet		PIA cable	NA	NA
13	Transmission to frame Lynx LCS 14E, 04076	4	14 inch	Metal	Eyelet		PIA cable	NA	NA	Eyelet		PIA cable	NA	NA
14	Hot in run & start relay 1 to HR&S fuse block 2	12	2	Red	Amp	1/4	180351-2	S/A cir 11	30	Ring	1/4	32703	NA	NA
15	Ignition switch relay2 to ground	16	1	Black	Push On	1/4	31073	S/A cir 10	86	Spade	#8	31721	47362	1
16	"Start" switch to ignition switch relay 2	16	6	Red	Push On	1/4	31090	44386	LOAD	Push on	1/4 16 Ga	31073	S/A cir 10	85
17	Hot in run & start relay 1 to ground	16	1	Black	Push On	1/4	31073	S/A cir 14	86	Spade	#8	31721	S/A cir 15	NA
18	Ignition "ON OFF" to "Start" switch	16	1	Red	Push On	1/4	31090	S/A cir 16	1	Push on	1/4	31090	44386	1
19	Key on/off switch to ignition switch	16	2	Red	Ring	#8	31709	NA	1	Push on	1/4	31090	S/Acir 18	2
20	HAFB 1 to key on/off switch	16	6	Red	Insl Push on	F16 Ga	31713	FB 1	Fuse 11	Ring	16 Ga #8	31709	switch	input
21	Ignition switch relay 2 to inline fuse 5 amp	16	2	Red	AMP	16 Ga	31073	S/A cir 10	87	Push on	16 Ga	31037	S/A cir 46	N start sw
22	combined with circuit 21													
23	combined with circuit 24													
24	In line fuse 5 amp to neutral start switch	16	5	Red	Push On	16 Ga	31037	S/A cir 46	N start sw	Butt step	16 Ga to 20 Ga	38050	NA	NA
25	Neutral start switch to starter relay 3	16	5	Red	Butt Step	16 Ga to 20 Ga	38050	NA	NA	Push on	1/4	31713	NA hvyduty	85
26	Starter relay 3 to ground	16	1	Black	Push On	1/4	31073	S/A cir 25		Spade	#8	31721	S/A cir 15	3
27	Ignition sw relay 2 to HIR relay #12 pin 30	10	2	Red	AMP	10 Ga	180351-2	S/A cir 10	87A	Ring	1/4 stud 10 Ga	32703	FB 3	NA
28	Fuel pump relay #11 to fuel pump	16	5	Red	AMP	F16 Ga	31073	S/A cir 187	87	Butt step	16 Ga to 20 Ga	38058	NA	NA
29	Fuel pump connector to ground	16	6	Black	Butt Step	16 Ga to 20 Ga	38058	NA	NA	Spade	#8 stud 16 Ga	31721	S/A cir 15	4
30	Starter relay 3 to starter solenoid	8	9	Red	Ring	(M5) # 10	33501	NA HEAVY	NA DUTY	Ring	(M5) # 10	33501	NA	NA
31	Fuel tank mount to frame	4	1	Metal	Ring	3/8	PIA cable	NA	NA	Ring	3/8	PIA Cable		
32	Maxi fuse block fuse 18 to voltage regulator	12	8	Red	Butt	12 Ga	32960	FB 1	Fuse 18	Butt	12 Ga	32960	NA	NA
33	HIR FB 3 to voltage regulartor	16	8	Red	Insl Push on	F 16 Ga	31713	FB 3	Fuse 31	Butt step	16 Ga to 18 Ga	38058	donor veh	F
34	Maxi fuse block fuse 2 to primary cooling fan relay 4	12	2	Red	Butt Step	12 Ga to 14 Ga	38051	FB 1	Fuse 2	Push on	1/4 12 Ga	31073	75280	30
35	Primary cooling fan relay 4 to cooling fan motor 1	12	11	Red	Push On	1/4-12 Ga	180351-2	S/A cir 34	87	In line butt	14 GA	31771	NA	NA
36	Circuit 37 to primary cooling fan relay 4	16	2	Red	Solder to circuit 37			TAPE	NA	Push on	16 Ga	31073	S/A cir 34	85
37	Hot in run FB 3 to secondary cooling fan relay 5	16	2	Red	Insl Push on	F16Ga	31713	FB 3	Fuse 32	Push on	16	31073	75280	85
38	Secondary cooling fan relay 5 to motor 2	12	11	Red	Amp	12 Ga	180351-2	S/A cir 37	87	In line butt	12 GA	32960	NA	NA

Example of wiring diagram used to design/develop the electrical system.

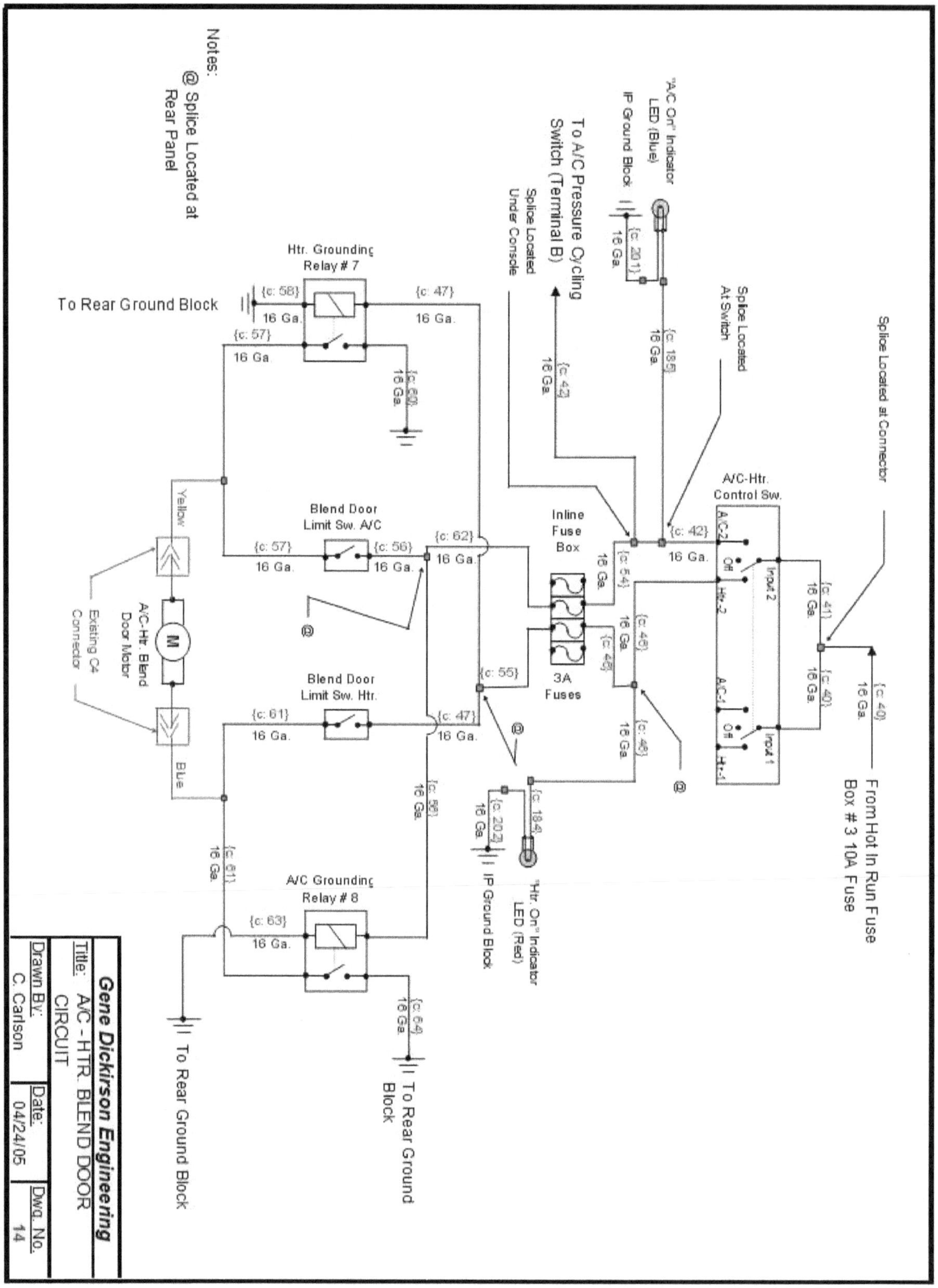

Engine Air Inlet Duct

The engine air inlet duct was designed on a full size layout then a CAD solid model was prepared. The CAD database was sent to Method Industries where it was fabricated using the same techniques as the body panels and the instrument panel. The duct is attached to the throttle body and air cleaner outlet using stainless steel worm gear clamps and custom flex couplings fabricated from sheet rubber. The duct also houses the engine air inlet temperature sensor.

Engine air inlet duct was fabricated in fiberglass and vinyl ester resin.

Engine Coolant Overflow Catch Bottle

The coolant overflow catch bottle is a medium size aftermarket unit. It is mounted on a rigid bracket that is attached to the front lower radiator support at a height that is well below the coolant surge tank to ensure proper operation.

Coolant overflow bottle is mounted at the appropriate height below the reservoir.

Engine Coolant Surge Tank

The engine coolant surge tank is fabricated from 5052 H aluminum sheets 2.29 mm (.09 inch) thick. The filler neck is a machined aluminum component supplied by Speedway Motors. The assembly is welded together. It is attached to the A/C evaporator assembly with screws.

Engine coolant surge tank was fabricated in aluminum.

Engine Fuel Injector Covers

The plastic covers that are used to hide the fuel injectors were modified by filling the graphics and holes with epoxy putty and smoothing. Flame Spray Coating Co, Inc. spayed them with molten aluminum with the intent of achieving the appearance of cast aluminum. The color of the sprayed aluminum did not match the color of the nearby engine castings so it was decided to paint the covers black to match the other black parts in the engine compartment.

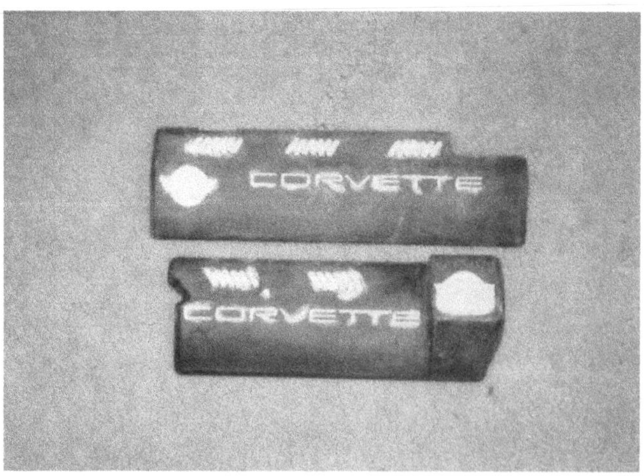

Engine fuel injector covers were modified by filling with epoxy putty and painted.

Exhaust System

The new exhaust system was designed and developed on the frame/ drive train and on the full size layouts. The precise locations of the ends of the donor exhaust pipes were found thru measurements to the new frame. The ends of the donor exhaust pipes (downstream of the catalytic converters) were the beginning of the new exhaust system. A custom "Siamese" muffler was designed and CAD detail drawings were prepared.

The detail drawings were sent to Spin Tech Performance Mufflers in Riverside, California who fabricated the muffler and shipped it to our shop. The tail pipes were designed then mocked up using 3" diameter plastic tubing and plywood and polystyrene foam. The mock up revealed areas that could be improved and changes were made to the design.

Tail pipes being mocked up in plastic tubing, plywood and foam.

CAD detail drawings were prepared of the tail pipes and the drawings were sent out for quotes to several firms. Based upon feedback comments from some of the firms that quoted on the tail pipes the design was again revised to facilitate fabrication. Stainless Works in Chagrin Falls, Ohio mandrel bent the stainless steel tubing and shipped the tail pipes to our shop. The tail pipe tips were designed by Craig Sandvig to match the opening on the lower area of the rear deck.

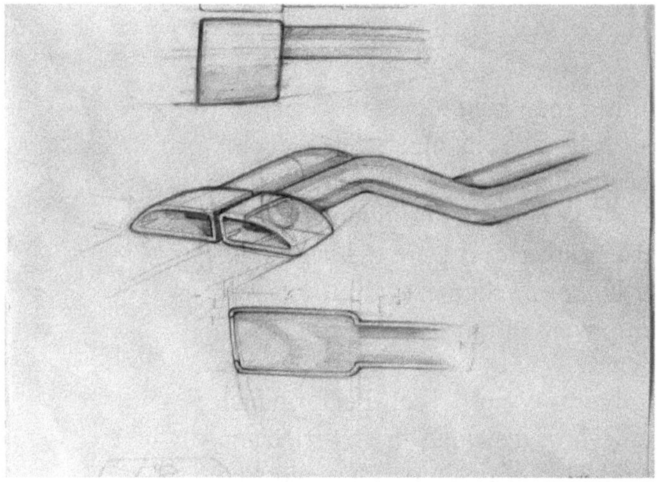

Craig Sandvig's sketch for beginning the design of the tail pipe tips.

They tips were mocked up using corrugated paperboard and plywood. The mock-ups were disassembled and their flat patterns along with CAD details were used by Stewart Steel Specialties in Farmington Hills, Michigan to fabricate the tips in aluminum. A stainless steel "adapter plate" was designed and fabricated. The stainless steel plate was welded to the stainless steel tail pipes and the aluminum tips were bolted to the plate using stainless steel fasteners. The pipes are attached to the muffler using stainless steel "T bolt" clamps. A bracket assembly fabricated from stainless steel is used to precisely position the rear end of the muffler laterally by tying it to the aluminum beam that connects the transmission to the differential. A stainless steel draw bar coil spring supports the tail pipes and tip assembly from the frame rear cross member. A set of draw bar coil springs support the tail pipes to the differential.

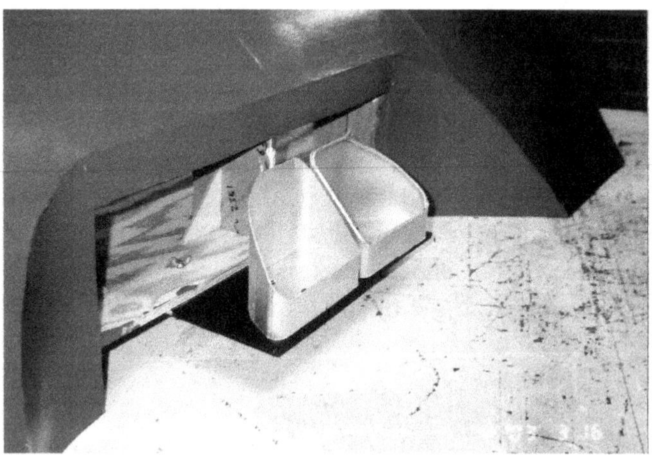

Tail pipe tips mocked up in chipboard mounted on the foam mock up of the vehicle tail.

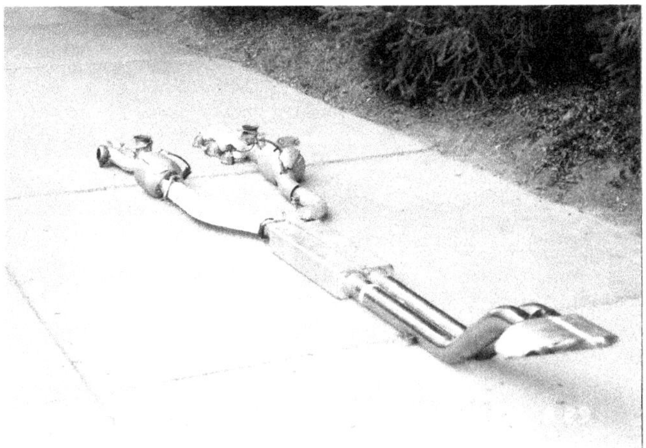

Complete exhaust system - front to rear.

Fuel Tank

The fuel tank was designed to fit between the rear frame rails, the rear cross member, the rear panel of the storage compartment, the rear deck support structure and the rear suspension support bracket. An expert on fuel tanks was consulted on several aspects of the tank including material, venting system, anti slosh reservoir, fuel level indicator sending unit and others. The 60 L (16 US gal) tank was designed to utilize a Corvette fuel pump/fuel level sending unit/filler neck/cap module. The new tank package required the fuel level float arm to be modified to fit inside the tank. The tank is fabricated from 5052 H aluminum sheet 2.3 mm (.09 inch) thick welded together. The aluminum tank mounts are welded to the tank and the mounts are attached to the frame via rubber isolators. An electrical ground strap connects the tank to the vehicle frame. The tank includes an internal reservoir that surrounds the fuel pump and maintains a small quantity of fuel to minimize the risk of fuel starvation during cornering/braking when the tank fuel level is very low. The fuel tank was leak tested after welding by filling it with water then pressurizing it with air to represent actual operating conditions.

Fuel tank being filled with water prior to leak test.

Fuel tank being leak tested with air pressure on top of water.

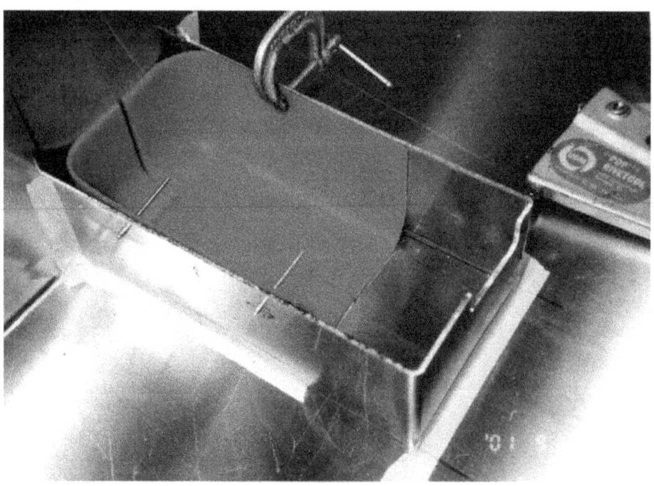

Fuel tank has an internal reservoir to ensure fuel delivery during turns with low fuel in tank

Headlamps

The 5.75-inch headlamp housings are the type used on motorcycles. The housings have adjustable mounting brackets that are attached directly to the steel hood support structure. The headlamp lenses are by Adjure with the 3K Diamond facet pattern. The metal closer rings have shields added to the front to prevent see thru between the headlamps and the fiberglass hood.

The headlamps are mounted directly to the hood support structure.

Heat Shields

Two sheet metal shields were designed and fabricated to minimize radiant heat transfer from the catalytic converters. One is fitted on the passenger side floor panel/ toe board and it is used to prevent overheating the passenger's feet. The other shield is fitted to the frame on the driver's side and it prevents the brake fluid from boiling.

Sheet metal shields passenger floor from catalytic converter heat.

Driver side heat shield protects rear brake line.

Horns

The set of horns are after market units- one high pitch and one low pitch. They are mounted on the frame just in front of the front anti sway bar.

Horns are mounted below the frame rails.

Inspection and VIN Tag

The GDT Speedster is titled by the State of Michigan as a 1994 Corvette. It carries the same vehicle Identification Number (VIN) as the donor vehicle. The vehicle was inspected by Plymouth Township police officer Steve Coffell on 28 August 2005 who verified the engine VIN matched the VIN listed on the title. Michigan Secretary of State officer Richard Zavitz installed a Replacement VIN tag on 20 October 2005.

Plymouth Twp. police officer Coffell inspecting the vehicle.

Michigan Secretary of State officer Zavitz affixing the VIN tag to the door pillar.

Laser Etched Components

The aluminum wheel center caps, horn button and fuel filler door were laser etched with the team logo. The etching was done by Great Lakes Laser Services based upon engineering drawings we provided to the firm.

Team logo laser etched on wheel caps, horn button and fuel filler door.

License Plate Frame Assembly

The license plate/ frame assembly is attached to the lower side of the rear deck using two knobs with 1/4-20 studs on them. The assembly is easily removable by unscrewing the knobs. The electrical connectors for the illumination lamp are integrated into the assembly and do not require any extra steps to connect/disconnect the circuit. The assembly is easy to remove and reinstall for displaying the vehicle at car shows. The assembly also incorporates a metal tab covered with felt that serves to dampen vibration. The felt protects the paint from being scratched.

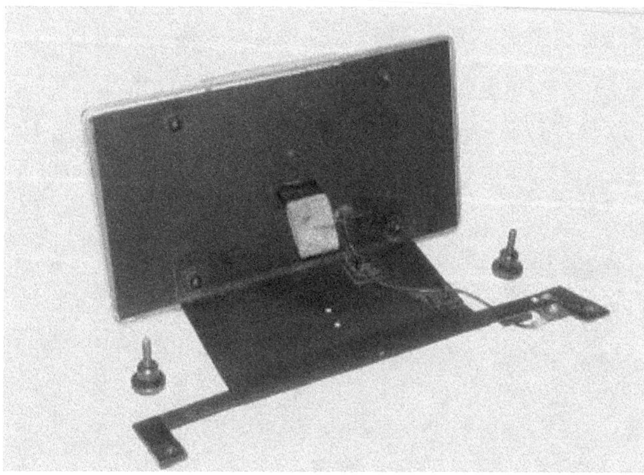

Rear Wheel Toe Adjusters

A unique set of adjusters were designed and fabricated for adjusting the toe angle of the rear wheels. The adjusters were designed on a full size layout. They use racecar double threaded knurled aluminum tubes and Heim joints supplied by Speedway Motors. The outer Heim joint ends are attached to the tapered holes in the rear knuckles using adapters supplied by Vette Brakes. The central mounting bracket is fabricated from welded steel plate. It is attached to the differential at the same location as the 1994 Corvette.

Unique rear wheel toe adjusters use racecar components.

Rear Storage Compartment

The rear storage compartment contains the battery on the driver side and the power train control module on the passenger side below its floor. The mounting panel for all of the major electrical components-relays, fuses and the On Board Diagnostics port- is housed on the rear wall of the storage compartment. The electrical component-mounting panel is fitted with a steel cover that has an electrical insulator made of acrylic sheet attached to its inner wall. A handle is attached to the top of the electrical panel cover to facilitate removal/ installation. Also the cover has three knobs to attach it to the rear wall of the storage compartment. The knobs eliminate the need for tools open the electrical panel to service the fuses.

Stewart Steel Specialties fabricated the storage compartment steel components using CAD details. It is a welded module and it is attached to the rear frame rails and rear bulkhead by ten M6 hex head cap screws.

Rear storage compartment side and rear walls being developed.

Cover for electrical panel can be easily opened by removing three knobs

Rear View Mirrors

The outside rear view mirrors are the type used on motorcycles. Their mounting rods have been shortened to fit the styling theme. They mount to the windshield support structure with M6 cap screws. Custom bezels were designed and fabricated in aluminum to cover the mounting screws.

The inside rear view mirror is by Billet Specialties, Inc. The mounting rod is fabricated in stainless steel. The mounting button is adhesively bonded to the windshield.

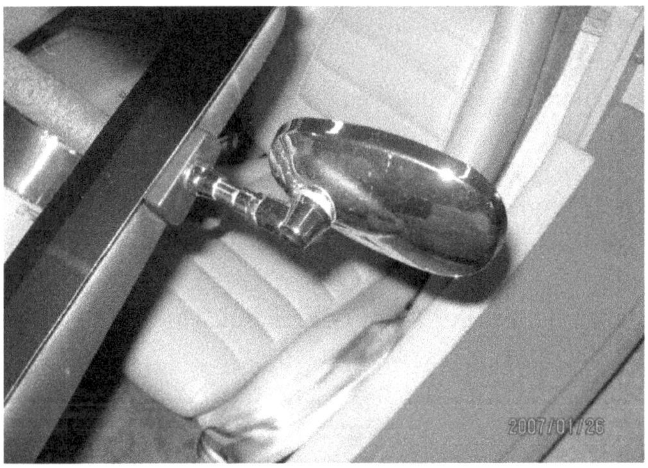

The outside rear view mirrors are modified motorcycle type.

The inside rear view mirror features a custom stainless steel mounting arm.

Ride Height

After the body panels were fitted a ride height study was conducted. The ride height of the vehicle -that is the distance between the vehicle frame and the ground-was intended to be the same as the 1994 Corvette. This height also controls the clearance between the tires and the wheel well flanges. In order to achieve the height objective it was planned to have the same weight as the 1994 Corvette so that the springs would deflect the same amount. The weight of the completed vehicle was projected based upon the known weight of many of the components. Every component except the engine and transmission had been weighed before it was assembled onto the vehicle. The weights of the engine and transmission were found in the book "Cor-vette Specs" by Mike Antonick, published by Michael Bruce Associates, Inc., Powell, Ohio. The weight of each component not yet completed was estimated based upon similar parts or engineering calculations. The projections indicated the vehicle would weigh approximately 1350 kg (3000 lb) versus the target weight of 1507 kg (3350 lb). A study was conducted to project the final ride height of the vehicle versus weight. The height of the partially completed vehicle was measured at each of the four corners of the frame (between the frame and the floor). A series of sand bags and iron lifting weights were placed in the cockpit in increments of approximately 22 kg (50 lb). After each weight increment was added the heights were measured. The heights were again measured as each increment of weight was removed. The height measurements were averaged and plotted against the weight increments.

The results indicated the target height would be met if approximately 135 kg (300 lb) of ballast were added to the vehicle. Research was conducted on several alternative ways of adding ballast to the vehicle including lead shot, lead ingots, iron lifting weights and steel shot. Steel shot was selected because the cost per unit volume was the lowest and the shot could be placed "out of sight" in the frame rails. Steel shot was purchased from Metaltec Steel Abrasive Co. in Canton, Michigan and approximately 135 kg (300 lb) was added to the frame rails by drilling 25 mm (1 inch) diameter holes in the tops of the rails. The holes were covered with sheet metal squares and painted after the shot was added.

After the vehicle was complete except for upholstery it was weighed at a commercial scale. The weight was 1503 kg (3340 lb).

It was found that driving the vehicle caused it to settle on its suspension system by a few millimeters. The actual vehicle height was found to be approximately 10 mm (.39 inch) lower than the target. This resulted in tire rub against the front and rear wheel well flanges under extreme wheel jounce conditions. It was necessary to remove all of the ballast from the frame in order to achieve adequate clearance between the front tires and the front wheel well flanges. In the rear, spring bolts 25 mm (1 inch) longer than standard were used to obtain the desired ride height.

Security Key

The original assumptions were to have no key for a "no clutter" appearance. Chuck Carlson recommended a security key be added for the owner's peace of mind. A pocket was designed into the end of the instrument panel on the driver's side to house the key that is visible only when the door is open.

The security key is housed in a pocket molded into the instrument panel.

Service Manual

A service manual was compiled to accompany the vehicle. It included the procedures for removing and reinstalling components unique to the vehicle. It included photos and text as well a complete set of electrical system schematics.

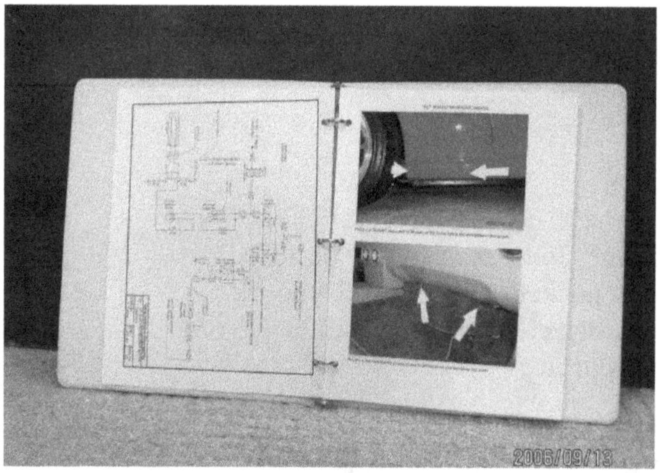

The Service Manual provides procedures for removing/reinstalling components.

Steering Column and Steering Wheel

The steering column was designed on the layout and developed using a wood mock up in the vehicle. The steering wheel location was the starting point of the process of designing the steering column. The steering wheel was located on the layout in the same relationship to the driver as in the 1994 Corvette, i.e. on the same plan view centerline as the driver seat, the same distance up and forward from the H Point and at the same angle. The column mounting brackets were developed to attach the column to the cross car beam and to the pocket in the dash panel. The brackets were designed and CAD details were prepared of all of the steering column components including the steering shaft. The steering shaft has a unique cross-section, i.e. different than the standard shafts used in aftermarket steering columns, to mate to the Corvette universal joint. The CAD details were sent to the ididit factory in Tecumseh, Michigan. The column, shaft and head were fabricated less the turn signal switch and wiring and returned to us. We fabricated the mounting brackets and built a weld fixture to hold them in the correct relationship to the column while Starlight Welding welded them. The column was then returned to ididit for final assembly of the electrical components and the shaft bearing.

The steering wheel was custom fabricated by Lecarra using leather supplied by our team. The horn button features the laser etched team logo.

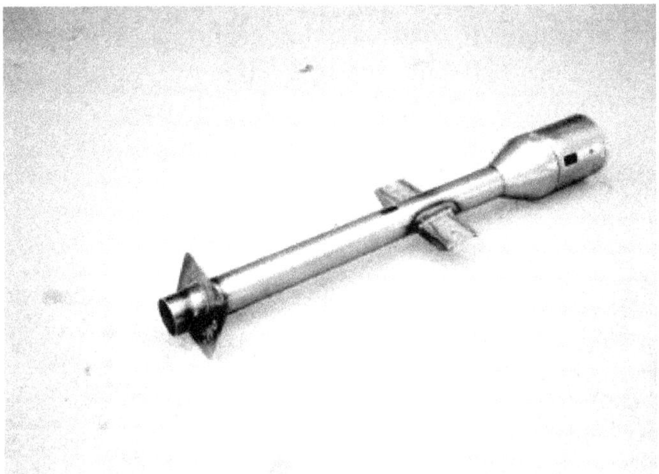

Steering column with mounting brackets welded in place.

Turn Signals

The front turn signals are the type used on motorcycles. They are mounted on the frame directly behind the grill. They are visible thru the grill slots when flashing.

Turn signals are mounted to the frame behind the grill.

Under Body Seat Support Straps

Based upon the recommendation of Musa Azzouz the seats are supported with underbody straps that are attached to the rocker panel mounting rails. The straps are intended to distribute the seat loads to the frame rails and minimize floor pan "oil canning". The straps are fabricated from 3.1 x 50.8 mm (.125 x 2 inch) steel. Each strap has three segments with bolted joints near the center of the vehicle to permit service to the exhaust system without the need to loosen the seat mounting nuts.

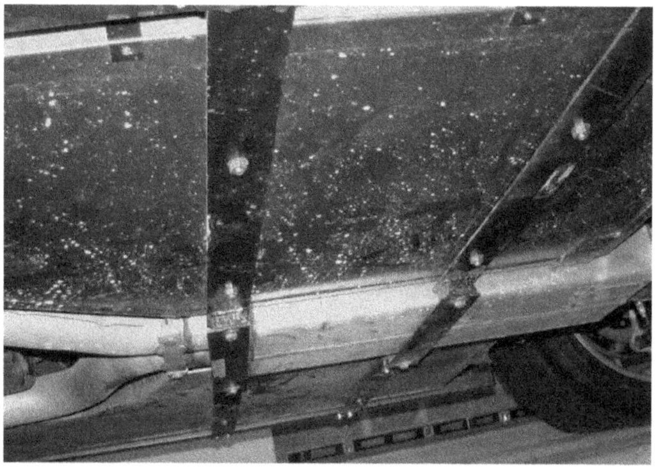

Seat support straps transfer loads to side rails.

Wheel and Tire Size Development

A study was conducted to determine the optimum sizes for the tires and wheels. The objective of the tires and wheels was to maximize tire width for appearance, use 18-inch diameter wheels for appearance and maintain the same normal tire diameter as the 1994 Corvette to minimize speedometer calibration issues. Chuck prepared a spreadsheet listing the wheel and tire sizes used on the 1994 Corvette. Both standard and optional sizes were listed. With the tire sizes known the outside diameters and widths were calculated.

The static loaded radius of the tires was determined by measuring several production C4 Corvettes on used car lots. The static loaded radius is the distance from the center of the wheel to the ground and it is smaller than the normal radius of the tire because the tire bulges slightly where it sits on the ground. Another objective was to maintain the same clearance between the inside faces of the tires and the suspension components/frame rails as the donor vehicle. Corvette track widths and wheel off sets were determined thru published data. The spacing between hubs was determined by subtracting the track widths from the wheel off sets. Revolutions per mile were also calculated for every study alternative. Thru research of tire sizes available with high-speed ratings several alternative tires were evaluated. The tires selected based upon the study were Michelin Pilot Sport, 285/ 35 ZR 18 front and 335/30 ZR 18 rear.

The three-piece wheels selected were fabricated by Fikse U.S.A. using off the shelf components with the off set and widths we specified. The polished forged aluminum wheels feature the Fikse Mach V spider and center caps laser etched with the team logo.

Windshield Wiper

The windshield wiper motor, blade and arm system is an aftermarket unit. It was included to ensure compliance with the laws of many states. It is mounted to the upper edge of the windshield with a custom designed and fabricated bracket. The electrical connection is made into a jack located in a pocket in the end of the instrument panel next to the security key pocket. The washer system is an aftermarket unit that utilizes a vinyl bag reservoir with a built in pump. The switch for the washer is attached to the wiper motor.

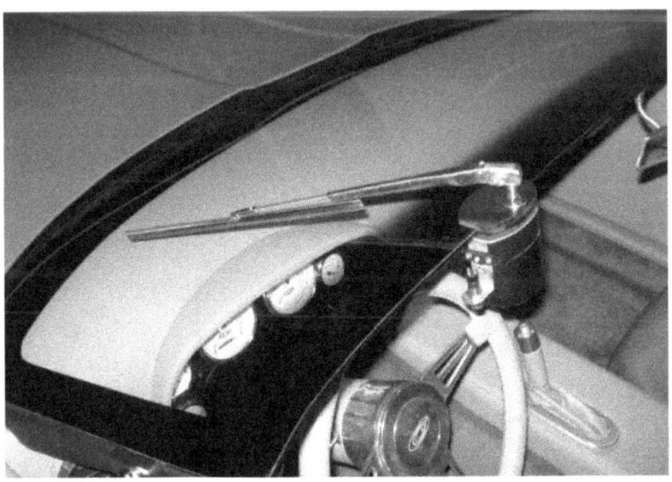

Windshield wiper is mounted to upper edge of windshield.

Team Logo

The team logo designed by Craig Sandvig.

12

First Start

The first attempt to start the vehicle on 13 May 2005 was unsuccessful and the cause was found to be a faulty fuel pump. The second attempt on 20 May was successful in that the engine started however a concern was identified. The concern was related to one of the wires being connected to the incorrect fuse block. The wire was connected to the proper fuse block and the engine started and ran as intended. The first test drive was conducted on 24 May 2005. The vehicle was driven to Belle Tire in Canton, MI on 31 May 2005 and the wheels were aligned to 1994 Corvette specifications. A series of road test drives then began and stretched out over several weeks and approximately 500 miles. A few small concerns were identified and corrected.

The first start of the engine was a gratifying experience.

The first drive confirmed that more than 2,000 parts functioned as a system.

Driving the GDT Speedster

Driving the vehicle is a lot of fun. To open the door one simply reaches in to the passenger compartment and pulls the door handle. The seat belts are manually adjusted to fit the driver and passenger. They are not equipped with automatic retractors making the "getting ready" experience more interesting.

To start the engine the driver simply pushes the ignition rocker switch to the "on" position and the adjacent indicator lights up green. Pushing the red rocker switch starts the engine. The sound of the exhaust system is throaty and is at the expected noise level. The transmission selector is easy to use and the gear selected is shown on the PRNDL gauge in front of the shift lever.

The car accelerates in low gear so fast that it is a challenge for the driver to prevent wheel spin. Heads turn as the car moves thru traffic. It is common for people in other cars to pull along side and ask, "What kind of car is that? I have never seen one. It is beautiful."

On the open road the car handles as expected as far as acceleration, turning and stopping are concerned. One pleasant surprise is the noise level is low. It is easy to carry on a conversation between the driver and passenger in a normal tone of voice. One of the factors that make the car quiet at road speeds is the lack of a windshield frame. The air slides over the glass and has no resistance when it reaches the edge.

Driving the GDT Speedster is an exhilarating experience.

Intentionally Blank

13

Completed Vehicle

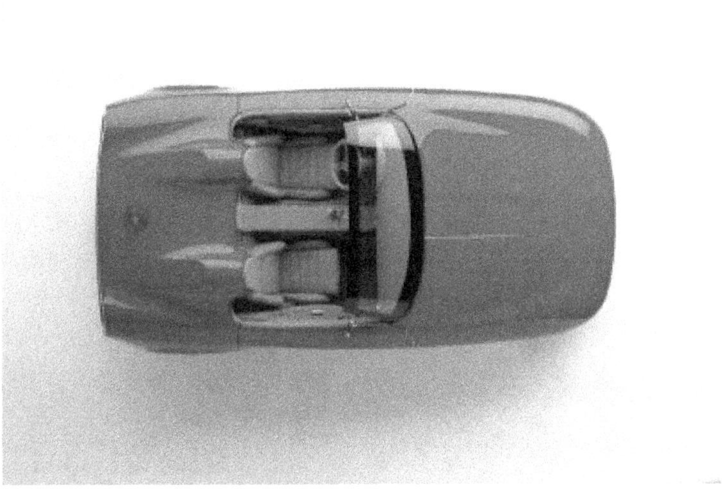

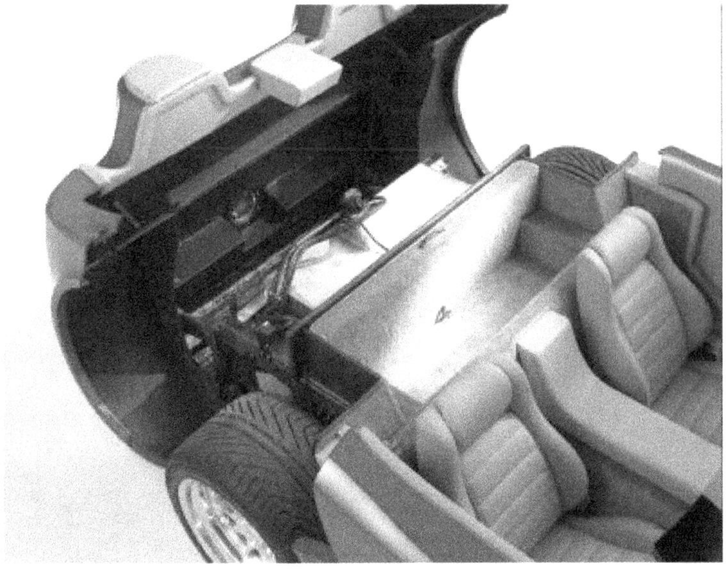

UQX 069

14

Publicity Campaign

As soon as the vehicle was completed a publicity campaign was launched. The first step was to have the vehicle professionally photographed at Acme Photo Works in Highland Park, Michigan on 22 and 23 December 2005. The images served as the key element for the web site we developed. The web site contained many of the same images as in this book. The web site text was written to explain how and why the vehicle was built. A video was produced at the suggestion of the web site compiler James Paul Martin of IV Productions. The video showed the vehicle being driven and therefore verified the vehicle was real, not just a mock up. The web site was launched on 28 January 2006.

The web site was the source of many articles on the vehicle.

The next step was to begin contacting the editors and writers at various newspapers and magazines in person, by telephone and by e-mail to introduce the GDT Speedster. Early in the publicity campaign, at the request of several editors we added a page to the web site that was only available to the media. The special page contained high-resolution images that the publications could access with a password. We also prepared a press release that we provided to numerous publications and other web sites. The press release essentially summarized the information that was contained on the web site.

The vehicle was entered in only one car show. At the invitation of Bob Elton, the show organizer the vehicle was entered in the Rolling Sculpture car show in Ann Arbor, Michigan on 14 July 2006. The vehicle received the Car & Driver magazine Editor's Choice award at the show.

The only time the vehicle was shown it won a major award.

The publicity campaign was judged to be highly successful. The following list summarizes the results of the campaign:

Publication	**Date**
Ann Arbor News	January 29, 2006
U.S. Auto Scene Newspaper	February 13, 2006
Reinforced Plastics	March 2006
Automotive Design and Production	April 2006
Composites Technology	April 2006
Detroit News	April 5, 2006
Plymouth Observer News	April 20, 2006
Classic Driver – German Addition	May 2006
Livonia Observer News	May 11, 2006
Corvette Magazine	June 2006
Hemmings Muscle Machines	June 2006
Makes & Models Magazine	June 2006
Super Rod Magazine	July 2006
Ann Arbor News	July 15, 2006
Adhesive & Sealants Industry	September 2006
Detroit Free Press	September 14, 2006
Auto Leht (Estonia)	November 29, 2006
Autoweek	February 12, 2007
Car & Driver 2008 Dream Cars Daily Calendar	March 20, 2007

In addition to the above print articles, hundreds of web sites around the world posted articles on the GDT Speedster. As of October 2007 more than 47,000 pages have been viewed on the GDT Speedster web site. The GDT Speedster video posted on www.youtube.com has been viewed more than 2800 times since it was posted in early 2006.

Intentionally Blank

15

Sales Campaign

The original strategy for selling the vehicle was to conduct an auction on our web site. However after some research it was decided to auction the vehicle on eBay. This was because it was felt that eBay has several methods in place to screen the bidders and therefore minimize the risk of erroneous bids.

Beginning on 14 April 2006 a direct mail campaign was initiated. Letters with one or two of the magazines and newspapers that contained articles on the GDT Speedster were sent to high profile car collectors. Their names were compiled from articles in newspapers, magazines, books, word of mouth and television. Their addresses were found on the Internet. A total of 15 sets of information were sent. The feedback was basically positive however there were no offers generated by the letters.

The next step in the sales campaign was to place ads on two web sites: collectorcartraderonline.com and cars-on-line.com. The ads began on 15 and 16 May 2006 respectively. The ads described the vehicle with several images and words and linked the readers to our web site. The cars-on-line.com "slide show" ad was the one with the highest hits for several days as soon as it was posted. On our web site we stated that the eBay auction would run for seven days and would begin on 25 June 2006. We invited potential bidders to see and drive the vehicle before the auction was scheduled to begin. There were no takers to see and drive the car.

To increase the potential number of bidders, ads were placed in the July 2006 issues of Corvette & Chevy Trader and the duPont Registry that were on the newsstands on 01 and 02 June 2006 respectively. The Corvette & Chevy Trader ad was the 4-color small size type and it showed the ¾ rear view of the vehicle. The text listed the price at $999,999 due to a typographical error. It drew several calls. The duPont Registry ad was located in the EXOTICS section of the magazine and showed a ¾ rear view of the vehicle. The text referred the reader to our web site and no price was listed. All of the magazine ads were accompanied by similar ads on their respective web sites.

The eBay auction began on schedule and there were 4505 hits on the listing however there were no bids placed. In hindsight it was felt the minimum bid of $200,000 was too high to get the auction started.The web site was revised in July 2006 to invite offers on the vehicle.

A series of four ads were placed in Autoweek magazine beginning on 20 July 2006. The ads included a ¾ rear view image and the text simply listed the GDT Speedster web site and a phone number. The ads drew a few "tire kickers" and no offers. A second ad in the September 2006 issue of Corvette & Chevy Trader hit the newsstands on 4 August 2007. It featured a 4-color large photo ad with a ¾ rear view and referred the readers to the web site. No price was listed. The cars-on-line.com ad was renewed on 9 August 2006.

The October 2006 issue of the Robb Report included our ad. Its newsstand date was 20 September 2006. It featured a side view of the vehicle and the simple text "The only one." with our web address and a phone number. The ad ran again in the November 2006 issue.

The GDT Speedster was a feature vehicle in the RM Auction catalog for their 2006 Fall Classic Car Auction to be held in Novi, Michigan. The Speedster was also mentioned in the article on the RM Novi auction in the 14 September 2006 issue of the Detroit Free Press. The vehicle crossed the auction block on Saturday 23 September and again on Sunday.

It drew a high bid of $70,000 on Saturday and $60,000 on Sunday, which was below our reserve, and therefore it did not sell.

The first live auction in Novi, MI on 23 September 2006.

On 10 October 2006 the text of the cars-on-line.com ad was revised to list a price of $150,000 where previously no price was listed. On 12 October 2006 the SALE PRICE page on our web site was revised to list an asking price of $150,000.

Our ad was placed in the December 2006 issues of Corvette & Chevy Trader and its companion magazine Old Car Trader. The ads were both the large 4-color size showing the side view of the vehicle and listed a price of $150,000.

The RM Auction catalog for their Ft. Lauderdale, Florida Collector Car Auction was delivered in the mail on 20 January 2007. It included a normal size photo of the GDT Speedster along with a brief description.

The GDT Speedster was sold for $60,000 at the RM Auction in Ft. Lauderdale, Florida on 10 February 2007. The GDT Speedster was purchased by Mr. John O'Quinn, the noted attorney and automobile collector of Houston, Texas. Mr. O'Quinn plans to display the vehicle along with others from his vast collection in the museum he is planning to build in the near future.

Loading for the trip to Ft. Lauderdale.

In the preview area at the Ft. Lauderdale auction.

16

Cost

Cost, functionality and timing are the major elements of any engineering project. Cost was the most sensitive element on the GDT Speedster project. Costs were tracked closely and meticulous records were kept.

One of the major challenges on the project was identifying suppliers who were willing to provide their products or services at a reasonable price. Many of the suppliers we contacted were accustomed to working with the major automobile manufactures and their prices were based upon their experiences with OEM customers.

In order to ensure we obtained the best value for the products we bought we issued a formal Request for Quote on nearly every major fabricated item we purchased for the project. The RfQ's were accompanied by an engineering drawing and written specification. The drawings and specifications minimized misunderstandings and unpleasant surprises. Frequently the potential suppliers would review the drawings and specifications and make recommendations for design changes that would result in significantly lower prices. We also found quotes that varied by as much as 10 times ($100 vs. $1000) on some products or services.

Even minor items were carefully shopped by obtaining the prices by from at least two potential sources.The project costs, EXCLUDING TEAM MEMBER HOURS are summarized in the following table:

Vehicle Specific Item	**Cost $**	**Project Related Item**	**Cost $**
Adhesives	137	Advertising (web site, magazine ads, photos)	4112
Body Panels	18532	Attorney fees (to form LLC)	2079
Brake & Fuel tubing	351	Auction fees	850
Carpet	66	Design/build materials/supplies	2016
Chevrolet Parts	452	Drafting materials	60
Donor Vehicle	5682	Freight (body panels from North Carolina)	744
Fasteners	353	Freight (vehicle to Ft. Lauderdale)	1300
Fiberglass materials	283	Michigan license & fees	463
Ford Parts	391	Office supplies	495
Frame Tubing	636	Postage & UPS fees	227
Gauges	922	Printing/ drawing reproductions	237
Head lamps	69	Travel expenses	1604
Leather	652	Vehicle Insurance	5642
Shift Knob (Machine Shop)	150	**TOTAL**	**19829**
Metal Bending	900		
Muffler	125		
Paint	13068		
PCM reprogram/wire harness	889		
Seat Belts	165	Power and hand tools and shop supplies	3032
Seats	1012		
Steel Sheet	402		
Steering Column	386		
Steering Wheel	149		
Tail Lamps	156		
Tail Pipes (stainless tubes bent)	214		
Upholstery labor	3934		
Welding	2197		
Wheels & Tires	5859		
Windshield(s)	554		
Wiring	801		
Sub Total	59487		
Other Parts, etc.	6877		
TOTAL	66364		

17

Frequently Asked Questions

The following numbers are the answers to frequently asked questions related to the GDT Speedster:

13, 127	Team members' hours devoted to the project
5.7	Years elapsed – beginning to completion of the build
1,997	Metal and fiberglass components fabricated for the vehicle
2,129	Fasteners utilized
61.4	Welding hours
719	CAD detail drawings completed
84	Full size pencil on vellum layouts completed
746	Individual purchases of components and materials
39	Laboratory tests conducted on adhesives and paints
1,080	Components in the vehicle electrical system
66,364	Build cost ($) excluding team members time/support
0	Cup holders

Intentionally Blank

18

Suppliers

The suppliers played an important part in making the GDT Speedster a success. The major suppliers are listed below:

Supplier	Product or Service	Telephone
Acme Photo Works	photos	313. 869. 9040
American Autowire	wiring components	800. 482. 9473
American Silkscreen	embroidery	248. 474. 1000
ArvinMeritor	gas springs	843. 464 5376
A.S.A.P. Machine Co.	shifter parts	734. 459. 2447
Auto Zone	parts	734. 721. 4199
Avery Dennison	tapes	800. 648. 7363
B&F Auto Supply NAPA	parts	734. 453. 7200
Baltek Corp.	fiber filler	201. 767. 7400
Belle Tire	tires & alignment	734. 844. 6001
Biggs Auto Renovation	paint	734. 459. 1064
Cap Plugs	sealing caps	888. 227. 5847
Cerullo Performance Seating	seats	909. 392. 5561
Classic Instruments	gauges	800. 575. 0461

Supplier	Product or Service	Telephone
Curbell	adhesives	800. 354. 4798
Devcon	adhesives	800. 933. 8266
Digital Design Inc.	surface modeling	586. 292. 4030
Electronic Connection	parts	734. 595. 6655
Extreme Adhesives	adhesives	800. 888. 4583
Family Brake & Repair	mechanical	734. 254. 9044
Federal Pipe & Supply	steel sheet & plate	313. 366. 3000
FedEx	shipping	313. 299. 8980
Fikse USA	wheels	253. 872. 3888
Flame Spray Coating Co, Inc	metal spray	800. 336. 8602
Freeman Mfg & Supply	sheet wax	800. 345. 9259
GKS Inspection Services	laser scanning	734. 953. 9696
Glassline	fiberglass	734. 453. 2728
Gordon Chevrolet	parts	734. 513. 1100
Great Lakes Laser Services	laser etching	248. 584. 1828
Hinkel Loctite	adhesives	860. 571. 5383
Howell Engine Developments	PCM/ wiring	810. 765. 5100
ididit	steering column	517. 424. 0577
Inline Tube	brake & fuel lines	800. 385. 9452
Interplastic Corp.	vinyl ester resin	651. 481. 6860
ITW Dynotech	drive shaft	248. 643. 2800
IV Productions, LLC	web site	734. 525. 3115
John Lyons	fittings	734. 591. 0606
Kraft Automotive Group	leather	517. 545. 7500
Laird's Auto Glass & Trim	windshield & trim	734. 453. 2599
Lou LaRiche Chevrolet	frame chart	800. 335. 5335
Lecarra	steering wheel	800. 432. 8170
Les Stanford Chevrolet	frame check	313. 565. 6000
Metaltec Steel Abrasive	ballast	734. 459. 7900
Method Industries	body panels	727. 798. 6587
Michigan Corvette Recyclers	parts	800. 533. 4650
Mobil Trim Shop	upholstery	734. 261. 1156
Motor City Fasteners	fasteners	248. 399. 2830

Supplier	Product or Service	Telephone
Murray's Auto Parts	parts	734. 459. 6200
Norma Products US	clamps	248. 668. 9510
Painters Supply	paint	734. 455. 4100
Plexus	adhesives	800. 851. 6692
Production Tool Supply	shop supplies	800. 236. 6020
Pro-Tec Products	sealing caps	800. 847. 0486
Ramchargers	parts	734. 591. 3224
Reid Tool Supply	hardware	800. 253. 0421
Rock Shoppe	weight scales	734. 455. 5560
S&W Pro Hardware	hardware	734. 453. 1290
Speedway Motors	suspension parts	402. 323. 3200
Spin Tech Performance Mufflers	mufflers	909. 360. 2474
Ssnake-Oyl Products	seat belts	800. 284. 7777
Stainless Works	tail pipes	800. 878. 3635
Starlight Welding	welding	313. 533. 3462
Stewart Steel Specialties	sheet metal	248. 477. 0680
Suburban Bolt & Supply	fasteners	800. 753. 6993
Summit Racing	parts	800. 230. 3030
3M	tapes & adhesives	800. 942. 1978
Tradesman Fastener	fasteners	313. 255. 7100
Tinnerman Palnut	fasteners	248. 226. 5901
Trim-Lok	trim moldings	714. 562. 0500
Vette Brakes	suspension parts	727. 345. 5292
Waytek	wiring parts	800. 328. 2724
Weldtube	tubing	313. 584. 6500
Wes-Garde Inc.	switch pads	630. 653. 1460
F.B. Wright	rubber sheet	800. 482. 0439
Zip Products	pedal pads	800. 962. 9632

Intentionally Blank

19

Other Contributors

The following people or organizations contributed to the success of the project by providing words of encouragement, advice or technical assistance.

David Traver Adolphus
Ken Ahn
Curt Alloy
Bob Amano
Lee Anderson
Stephen K. Anderson
Richard Arrowood & Family
Virginia Arrowood
Mike Antonick
Gene & Charlene Axelrod
George Balachak
Matthew J. Baumlin
Jim Bechtel
Henry Beckwith
Robert P. Belletini
Chuck Bentley
Phil Berg
Jeff Berge
Gary Berlin
Randy Betki
Greg & Denise Bishop
Steve Bittner
Sara Black
Barry Blummel
Dave Boerger
Ron Breuhan
George Brinket
Jerry & Gill Brown
Susan Brown
Tony Bruscato
Mike Bullion
Jim Burger
John Busch
Fay Butler
Ken & Jane Callison
Ron Celano
Richard A. Cerullo
Joe Choroba
Brian Childs
Jon Churgay
Chris G. Clark
Bill Connell
Mike & Patricia Cops
Ron Covell
Jack Criswell

Charles J. Curmi
Reggie Czach
Bob Deachin
Jim Dean
Steve DeRemer
Rudy DeSanto
David Dickirson & family
Donald Dickirson
Hattie Dickirson
Mark Dickirson & family
Mindy Dickirson
Christopher C. Dowdey
Richard Dubay
Dale Ducklow
Doug Dutton
Jim Dutton
Greg Dworzanin
Larry Edsall
Robert Elton
Don Elvy
Ralph Emmons
George Farley
F rank Feeney
Clark Ferguson
Al & Delores Fields
Lee Fields
John Filiss
Don Fink
Bob & Joan Florian
Henry Ford Community College
Del & Bonnie Fox
Levi Fox & family
Mike & Janet Fox
Terry & Julie Fredricks & family
Mark Freeland

Raymond Frost
Judy Gale
Mickey Galloway
Ben Garrett
Richard Glomb
Irena Granaas
Hank Granowicz
Dan Gray
Steve Gray
Tony Grech
Euge Greenstein
Les Greenup
Mike Greer
Bob Gritzinger
Mike Hammer
Leon & Sandy Hanson
Mel Harbaugh
Jason Haritou
Mike Harrigan
Roy Harris
Patty Hastie
Brian Hayes
Dennis & Linda Henderson
Jim Herrick
Charles Hicks
Darrel Higgens
George Hoffman
Jim & Mary Hollowood
Mike Hudson
Tom Hulls
Doug Imlach
Carl & Sue Ingrao
Amanda Jacob
Mark James
Gary Johns

Donald Johnson
Warren Johnson
Brad Kadrich
Paul Keberly
Mark Kelly
Jim Kempf
Ted Keranen
Guy Kenny
Mike Kettler
Ji- Hoon Kim
Kurt Kiser
Brian Kittle
Mike Kleeves
Gerald Klein
Doug & Kay Koch
Steve Kraft
Tim Kramer
Ron Kretchmer
Bill Krueger
John Krueger
Margus H. Kuuse
Tom Laird
Jay Lamm
Michael Lamm
Tom LaPointe
John Ledbetter
Ed Lewandowski
Mik Lisiecki
Eric Lux
John Lyon
Bob Malkowski
James Paul Martin
Joe Matyok
Bill & Betty Mayes
Catharine McKinney

Mike McKusick
Lynn McLean
Kevin & Pam McLoughlin
Will McWhirter
Henry Mehraban
Tiffany A. Mitchell
Joe Mooradian
Richard V. Munz
Ken & Diane Obudzinski
Jack O'Connell
Mike Ondayko
Ruth Page
David & Helen Palechek
Mike Papoulias
Ram Patel
Frank Pellegata
Jeff & Colleen Peterson
Fred Pilgrim
Don Plester
Bruce Pollock
Nick Pon
Mike Ramsey
Carol Rasak
Paul Rebman
Neil Ressler
Gene Rhodes
Rick Roberts
Bill & Kathy Rodreick
Bob & Barbara Rodreick
Ken Ross
Robert B. Ruokolainen
David Ryan
Greg Rygiel

Dan Sandel
Hylber Sandvig
Teresa Sasinowski
Nick Satterwaite
Christopher A. Sawyer
Tom & Farla Schiermeier & family
Fred Schroeder
Ed Scutchfield
Mary Sealhurst
Walt Sears
Ralph Skwarski
Bernard Smith
Bob Smith
Gene Smith
Jerry Smutek
Joe Sroka
Dick Stanhope
Terry Steller
Steve Stropes
John C. Stewart
Cristina Stuart
Dave Sulhan
Tony Swan
Mike Thompson
Harold Thorne
D. Mark Trostle
Jim & Jane Ann Turpen & family
Joe Uhl
Ray Vagnetti
Jim & Malisa Voss
Jan Wade
Roger Walker
Jeff & Michelle Warmolts

Marty Waterstraut
Ross Weaver
Gary Weber
Russ Wee
Matt Wegrzynowicz
Bob Wheaton
Mary Ann Wheeler
Mike Wheeler
John White
George Wiklund
Allan Lee Wilber & family
George T. Wilber & family
Harold Gene Wilber & family
Jim Walker Wilber & family
Mildred Wilber
Bill Williams
Pat Willey
Joe & Polly Winkle
Sharon Wood
Tom Wright
David Yezback
Dough Young
Bill Youskovich
Jeff & Karen Zokas

Intentionally Blank

20

Photo Credits

The studio photos were taken by Nathan Garcia assisted by Mike Merill at the Acme Photo Works studio in Highland Park, Michigan on 22 & 23 December 2005.

Phil Berg took the photos of the vehicle in motion. James Paul Martin took the video photo. Musa Azzouz, Chuck Carlson, Larry Ronzi, and Craig Sandvig provided their respective photos. Barbara and Gene Dickirson took the remainder of the photos.

Nathan Garcia (L) and Mike Merill (R) taking the glamour photos at Acme Photo Werks.

A tall stepladder was needed to get to crows nest for the overhead photos.

21

Specifications

Overall length	4110 mm (161.8 inches)
Overall height	1068 mm (42.0 inches)
Overall width	1897 mm (74.6 inches)
Front track	1502 mm (59.1 inches)
Rear track	1550 mm (61.0 inches)
Weight	1471 Kg (3244 lb)
Fuel tank capacity	60 L (16 gal US)
Leg room	1067 mm (42.0 inches)
Shoulder room	1460 mm (57.5 inches)
Engine	Chevrolet Corvette LT1
Engine power output	224 Kw (300 HP) @5000 RPM
Engine torque	461 Nm (340 ft lb) @ 3600 RPM
Transmission	Hydra-Matic 4L-60 E
Rear axel ratio	2.59:1
Transmission ratios	1st gear 3.06:1, 2nd gear 1.63:1, 3rd gear 1.0: 1, 4th gear 0.70:1

Intentionally Blank

22

Performance

The following performance data is estimated based upon published data for Chevrolet Corvettes equipped with LT 1 engines. The actual performance of the GDT Speedster has not been determined under controlled, normally accepted testing procedures.

Acceleration	5.7 seconds 0 to 96 km/h (0 to 60 mph)
Braking Distance	37.4 m from 96 km/h (123 ft from 60 mph)
Fuel Mileage	6.8 km/l (16 mpg)
Top Speed	273 km/h (170 mph)

Intentionally Blank

Index

www.ingramcontent.com/pod-product-compliance
Lightning Source LLC
LaVergne TN
LVHW082337200926
848481LV00012B/3366
9780615173269